Benchmark Models of Control System Design for Remotely Operated Vehicles

Cheng Siong Chin · Michael Wai Shing Lau

Benchmark Models of Control System Design for Remotely Operated Vehicles

Springer

Cheng Siong Chin
Faculty of Science,
Agriculture & Engineering
Newcastle University in Singapore
Singapore, Singapore

School of Automotive Engineering,
The State Key Laboratory of Mechanical
Transmissions, Chongqing Automotive
Collaborative Innovation Center
Chongqing University
Chongqing, China

Michael Wai Shing Lau
Faculty of Science,
Agriculture & Engineering
Newcastle University in Singapore
Singapore, Singapore

ISBN 978-981-15-6510-6 ISBN 978-981-15-6511-3 (eBook)
https://doi.org/10.1007/978-981-15-6511-3

This Springer imprint is published by the registered company Springer Nature Singapore Pte Ltd.
The registered company address is: 152 Beach Road, #21-01/04 Gateway East, Singapore 189721, Singapore

Preface

The modeling, simulation, and application on a marine vehicle such as an underwater robotic vehicle (URV) reflect the multi-disciplinary that unifies mathematical aspects from the various and separate engineering disciplines. URVs such as a remotely operated vehicle (ROV) are used for a wide range of applications, such as exploring the extreme depths of our ocean where the hard-wired link is still required. Most of the ROVs operate in an extreme environment with uncertainties in the model before control system design. However, the method involved extensive testing before the system model could be used for any control actions. It has been found that the range of errors could be extensive and uncertain in actual continuously varying conditions. It is important to address the problem of reliance on model testing using different modeling approaches. Hence, in this book, approaches such as WAMIT^{TM}, STAR-CCM+^{TM}, MATLAB^{TM}, and Simulink^{TM} are used to model the parameters of the ROV, resulting in a benchmark ROV model for control. The control systems such as proportional–integral–derivative (PID), backstepping, model predictive control, and fuzzy-logic control are simulated. The benchmark model of the ROV is given to allow researchers and students to explore and test different control schemes on the same platform.

This book has been prepared for the needs of those who seek an application on developing control systems for ROV where the model is not available during the initial design stage. Also, some models in the literature are difficult to understand and details on how to obtain the model parameters can be missing. It often impedes the process of control systems design if details of the ROV model are not available. The book can be used as a reference book for undergraduate and postgraduate, engineers, researchers, and lecturers in modeling and simulation for control using MATLAB^{TM} and Simulink^{TM} as the main software.

Singapore, Singapore

Cheng Siong Chin
Michael Wai Shing Lau

Acknowledgments

Finally, I wish to acknowledge my great debt to all those who have contributed to the development of modeling and simulation into the field it is today and specifically to the considerable help and education I have received from final-year project students, industry partners, and my colleagues. In particular, I would first like to thank, my late master thesis advisor: Prof. Neil Munro who had made control engineering alive by using computer-aided software. I had benefited from his knowledge in the robust control and the computer-aided control engineering. In addition, I would like to thank Prof. Tuncer Ören from the School of Information Technology and Engineering at the University of Ottawa, Canada for his motivation and advice. I have to thank my colleague, Prof. Caizhi Zhang for his kind support and opportunity at Chongqing University in China. Lastly, I would like to thank the editorial staff and reviewers for providing comments to the text. The MATLAB™ programming codes and Simulink™ block diagrams are provided in this book. I hope to encourage more "deep" collaborative research.

I would like to express my gratitude to my family. In particular, I would like to thank Irene (my wife), Giselle (late elder daughter), Millicent (my second daughter), Tessalyn (third daughter), and Alyssa (youngest daughter) for their support and understanding, without which the undertaking would not have been possible. I would like to share my joy and results with my strong and kind daughter, Giselle. Although she passed away at the age of four after a few consecutive open-heart surgeries, without her presence and attitude toward life, many things would not have happened. I have collected all my tears and preserved them into this book to continue my academic life and turn that sadness into strength.

Prof. Cheng Siong Chin, Ph.D., M.Sc., B.Eng.,
Dip.Mech. FHEA, FIMarEST, SMIEEE, Eur Ing, MIET, CEng

Associate Professor Michael Wai Shing Lau, Ph.D., M.Sc., B.Eng.

Outlines

The contents of the book are organized into four main chapters with programming codes. The main chapters include a section on the modeling of remotely operated vehicle (ROV), model-based simulation of ROV, and identifications of final benchmark ROV model for control systems design. These chapters contain applications using MATLAB™ and Simulink™ on various subsystems of the ROV. As it is not possible to cover all controllers in the control systems design chapter, the chapter, therefore, emphasizes a few control systems such as proportional–integral–derivative (PID) controller, backstepping controller, model predictive controller, and fuzzy-logic controller. The proposed benchmark model of the ROV is obtained from the computational methods using mainly the computational fluid dynamics software such as STAR-CCM+™. The simulated results are verified with the experimental tests in the water tank. The software adopted throughout the chapters is MATLAB™/Simulink™. The programming codes and block diagrams are available in the book.

The chapters are written in a coherent manner where each chapter is closely connected. This allows the readers to understand the entire process of modeling, simulation to control systems design.

- Chapter 1: Modeling of Remotely Operated Vehicle—provides a detailed model of the ROV;
- Chapter 2: Simulation of Remotely Operated Vehicle—gives the simulation of the model derived in Chap. 1;
- Chapter 3: Identification of Remotely Operated Vehicle and Benchmark ROV Model—presents the parameters of the ROV used in Chap. 2. The detailed steps to obtain the benchmark model of the ROV are provided;
- Chapter 4: Control Systems Design of Remotely Operated Vehicle—shows how to control the benchmark model of the ROV in Chap. 3.

Contents

Chapter 1
Modeling of Remotely Operated Vehicle

Abstract In this chapter, nonlinear dynamic model of an ROV with perturbation due to hydrodynamic forces is discussed. In general, there are two methods to derive the dynamic model: the Lagrange formulation using the energy method and Newtonian's formulation using the Newton's Second Law of Motion. The latter method gives an equation using the angular velocities about the body-fixed axes; unlike the generalized variables used in the Lagrange equation, the angular velocities cannot be integrated to obtain angular displacements about these axes and are therefore unsatisfactory to describe fully the orientation of a rigid body in three-dimensional (3D) space. However, the body-fixed angular velocities can be solved and transformed to give the orientation of the vehicle in 3D space using the Euler's angles. The Euler's transformation (or commonly known as kinematics equation) provides the relationship between the dynamics derived in the earth-fixed and the body-fixed coordinates. There are practical advantages to derive the dynamics in the body-fixed coordinate for the following reasons. First, most sensors and actuators mounted on the ROV measure the body parameters, such as the ROV velocities, and provide propelling forces. Secondly, the dynamic equation is inherently parameterized and hence less complicated because this parameterized form is not Euler's angle-dependent. Unfortunately, due to modeling uncertainties, the model derived by either method can be inadequate for control system design. These uncertainties are mainly due to the hydrodynamic forces which are experimentally and theoretically difficult to obtain. Nevertheless, for control design purposes, there is a need to consider a simplified nominal model with additive perturbation bounds that are accurate enough to represent the ROV dynamic behavior.

1.1 Dynamic Equation Using the Newtonian Method

The following assumptions can be made when deriving a typical ROV dynamic equation to simplify the effort in modeling:

(a) ROV is a rigid body and is fully submerged once in water;
(b) Water is assumed to be an ideal fluid that is incompressible, inviscid (frictionless), and irrotational;

C. S. Chin and M. W. S. Lau, *Benchmark Models of Control System Design for Remotely Operated Vehicles*, https://doi.org/10.1007/978-981-15-6511-3_1

Table 1.1 Notations used for ROV

DOF	Motion descriptions	Forces and moments	Linear and angular velocity	Positions and orientations
1	Motion in the *x*-direction (surge)	X	u	x
2	Motion in the *y*-direction (sway)	Y	v	y
3	Motion in the *z*-direction (heave)	Z	w	z
4	Rotation about *x*-axis (roll)	K	p	ϕ
5	Rotation about *y*-axis (pitch)	M	q	θ
6	Rotation about *z*-axis (yaw)	N	r	ψ

(c) ROV is slow-moving for operation such as pipeline inspection;
(d) The earth-fixed frame of reference is inertial;
(e) Disturbance due to wave is neglected as it is fully submerged and underwater current is treated as a persistent perturbation;
(f) Tether dynamics attached to the ROV is not initially modeled.

The notations (see Table 1.1) used for the ROV motions follow the Society of Naval Architects and Marine Engineers (SNAME).

A rigid body ROV dynamic equation is commonly expressed in the body-fixed frame since the control forces and measurement devices are easily and intuitively related to this body frame of reference. Using the Newtonian approach, the motion of a rigid body with respect to the body-fixed reference frame at the origin (see Fig. 1.1) is given by the following set of equations [1]:

$$\mathbf{M_{mass}}[\dot{\mathbf{v}}_1 + \mathbf{v}_2 \times \mathbf{v}_1 + \dot{\mathbf{v}}_2 \times \mathbf{r_G} + \mathbf{v}_2 \times (\mathbf{v}_2 \times \mathbf{r_G})] = \boldsymbol{\tau}_1 \tag{1.1}$$

$$\mathbf{I}\dot{\mathbf{v}}_2 + \mathbf{v}_2 \times (\mathbf{I}\,\mathbf{v}_2) + \mathbf{M_{mass}}\mathbf{r_G} \times (\dot{\mathbf{v}}_1 + \mathbf{v}_2 \times \mathbf{v}_1) = \boldsymbol{\tau}_2 \tag{1.2}$$

where $\mathbf{r_G} = [x_G\ y_G\ z_G]^{\mathrm{T}}$ is the location of the centre of gravity from a fixed point O,$\boldsymbol{\tau}_1 \in \Re^3$ and $\boldsymbol{\tau}_2 \in \Re^3$ are the external force and moment vector, respectively, $\mathbf{v}_1 = \begin{bmatrix} u\ v\ w \end{bmatrix}^{\mathrm{T}} \in \Re^3$ is the linear velocity vector, and $\mathbf{v}_2 = \begin{bmatrix} p\ q\ r \end{bmatrix}^{\mathrm{T}} \in \Re^3$ is the angular velocity vector. $\mathbf{M_{mass}} \in \Re^{3\times3}$ is the ROV mass matrix:

$$\mathbf{M_{mass}} = \begin{bmatrix} m & 0 & 0 \\ 0 & m & 0 \\ 0 & 0 & m \end{bmatrix} = m\mathbf{I}_{3\times3}; \tag{1.3}$$

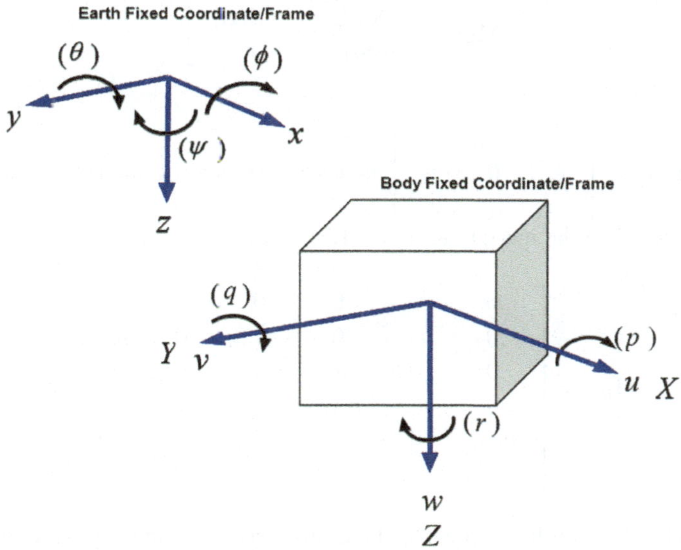

Fig. 1.1 Coordinate systems used in ROV

$\mathbf{I}_{3\times3}$ is the identity matrix and $\mathbf{I} \in \Re^{3\times3}$ is the inertia tensor defined at O:

$$\mathbf{I} = \begin{bmatrix} I_x & -I_{xy} & -I_{xz} \\ -I_{yx} & I_y & -I_{yz} \\ -I_{zx} & -I_{zy} & I_z \end{bmatrix}; \mathbf{I} = \mathbf{I}^{\mathrm{T}} > 0,\ \dot{\mathbf{I}} = 0 \tag{1.4}$$

I_x, I_y, and I_z are the moments of inertia about the X, Y, and Z axes, respectively, at O. As $\mathbf{I} = \mathbf{I}^{\mathrm{T}} > 0$, the cross products of inertia $I_{xy} = I_{yx}$, $I_{xz} = I_{zx}$, and $I_{yz} = I_{zy}$.

The rigid body equation can be expressed in matrix form as:

$$\mathbf{M_{RB}}\dot{\boldsymbol{\nu}} + \mathbf{C_{RB}}(\boldsymbol{\nu}) = \boldsymbol{\tau} \tag{1.5}$$

where $\mathbf{M_{RB}} \in \Re^{6\times6}$ is the mass-inertia matrix, $\mathbf{C_{RB}}(\boldsymbol{\nu}) \in \Re^{6\times6}$ is the Coriolis and centripetal matrix, $\boldsymbol{\tau} = \begin{bmatrix} \boldsymbol{\tau}_1 & \boldsymbol{\tau}_2 \end{bmatrix}^{\mathrm{T}} \in \Re^6$ is a vector of external forces and moments, and $\boldsymbol{\nu} = \begin{bmatrix} \boldsymbol{\nu}_1 & \boldsymbol{\nu}_2 \end{bmatrix}^{\mathrm{T}} \in \Re^6$ is the linear and angular velocity vector.

From (1.1) and (1.2) we have,

$$\mathbf{M_{RB}}\dot{\boldsymbol{\nu}} = \begin{bmatrix} \mathbf{M_{mass}}\dot{\boldsymbol{\nu}}_1 + m\dot{\boldsymbol{\nu}}_2 \times \mathbf{r_G} \\ \mathbf{I}\dot{\boldsymbol{\nu}}_2 + m\mathbf{r_G} \times \dot{\boldsymbol{\nu}}_1 \end{bmatrix} \tag{1.6}$$

Since $\mathbf{M_{RB}} = \mathbf{M_{RB}^{\mathrm{T}}} > 0$, it can be separated and written as:

$$\mathbf{M_{RB}} = \begin{bmatrix} \mathbf{M_{mass}} & -m\mathbf{S}(\mathbf{r_G}) \\ m\mathbf{S}(\mathbf{r_G}) & \mathbf{I} \end{bmatrix} \tag{1.7}$$

where $\mathbf{S}(\mathbf{r_G}) = \begin{bmatrix} 0 & -z_G & y_G \\ z_G & 0 & -x_G \\ -y_G & x_G & 0 \end{bmatrix}$ is a skew-symmetric matrix. Expanding the individual terms in the inertia matrix in (1.7) gives:

$$\mathbf{M_{RB}} = \begin{bmatrix} m & 0 & 0 & 0 & mz_G & -my_G \\ 0 & m & 0 & -mz_G & 0 & mx_G \\ 0 & 0 & m & my_G & -mx_G & 0 \\ 0 & -mz_G & my_G & I_x & -I_{xy} & -I_{xz} \\ mz_G & 0 & -mx_G & -I_{yx} & I_y & -I_{yz} \\ -my_G & mx_G & 0 & -I_{zx} & -I_{zy} & I_z \end{bmatrix} \tag{1.8}$$

Similarly, the Coriolis and centripetal terms, describing the angular motion of the ROV in (1.1) and (1.2), can be rewritten as:

$$\mathbf{C_{RB}}(\boldsymbol{\nu})\boldsymbol{\nu} = \begin{bmatrix} m[\boldsymbol{\nu}_2 \times \boldsymbol{\nu}_1 + \boldsymbol{\nu}_2 \times (\boldsymbol{\nu}_2 \times \mathbf{r_G})] \\ \boldsymbol{\nu}_2 \times (\mathbf{I}\boldsymbol{\nu}_2) + m\mathbf{r_G} \times (\boldsymbol{\nu}_2 \times \boldsymbol{\nu}_1) \end{bmatrix} \tag{1.9}$$

Using the results $\mathbf{a} \times \mathbf{b} \times \mathbf{c} = \mathbf{S}(\mathbf{a})\mathbf{S}(\mathbf{b})\mathbf{c}$ and $\mathbf{S}(\boldsymbol{\nu}_1)\boldsymbol{\nu}_1 = 0$, the preceding equation can be manipulated to give:

$$\mathbf{C_{RB}}(\boldsymbol{\nu})\boldsymbol{\nu} = \begin{bmatrix} -m\mathbf{S}(\boldsymbol{\nu}_1)\boldsymbol{\nu}_2 - m\mathbf{S}(\boldsymbol{\nu}_2)\mathbf{S}(\mathbf{r_G})\boldsymbol{\nu}_2 \\ -m\mathbf{S}(\boldsymbol{\nu}_1)\boldsymbol{\nu}_1 + m\mathbf{S}(\mathbf{r_G})\mathbf{S}(\boldsymbol{\nu}_2)\boldsymbol{\nu}_1 - \mathbf{S}(\mathbf{I}\boldsymbol{\nu}_2)\boldsymbol{\nu}_2 \end{bmatrix} \tag{1.10}$$

Separating $\boldsymbol{\nu}_1$ and $\boldsymbol{\nu}_2$ yields:

$$\mathbf{C_{RB}}(\boldsymbol{\nu}) = \begin{bmatrix} \mathbf{0}_{3\times3} & -m\mathbf{S}(\boldsymbol{\nu}_1) - m\mathbf{S}(\boldsymbol{\nu}_2)\mathbf{S}(\mathbf{r_G}) \\ -m\mathbf{S}(\boldsymbol{\nu}_1) + m\mathbf{S}(\mathbf{r_G})S(\boldsymbol{\nu}_2) & -\mathbf{S}(\mathbf{I}\boldsymbol{\nu}_2) \end{bmatrix} \tag{1.11}$$

Expanding the terms on the RHS gives:

$$\mathbf{C_{RB}}(\boldsymbol{\nu}) = \begin{bmatrix} \mathbf{0}_{3\times3} & \mathbf{C_{12}}(\boldsymbol{\nu}) \\ -\mathbf{C_{12}^T}(\boldsymbol{\nu}) & \mathbf{C_{22}}(\boldsymbol{\nu}) \end{bmatrix} \tag{1.12}$$

with

$$\mathbf{C_{12}}(\boldsymbol{\nu}) = \begin{bmatrix} m(y_G q + z_G r) & -m(x_G q - w) & -m(x_G r + v) \\ -m(y_G p + w) & m(z_G r + x_G p) & -m(y_G r - u) \\ -m(z_G p - v) & -m(z_G q + u) & m(x_G p + y_G q) \end{bmatrix} \tag{1.13}$$

$$\mathbf{C_{22}}(\boldsymbol{\nu}) = \begin{bmatrix} 0 & -I_{yz}q - I_{xz}p + I_z r & I_{yz}r + I_{xy}p - I_y q \\ I_{yz}q + I_{xz}p - I_z r & 0 & -I_{xz}r - I_{xy}q + I_x p \\ -I_{yz}r - I_{xy}p + I_y q & I_{xz}r + I_{xy}q - I_x p & 0 \end{bmatrix} \tag{1.14}$$

The external force and moment vector $\boldsymbol{\tau}$ includes the hydrodynamic forces and moments, $\boldsymbol{\tau_H}$ due to damping and $\boldsymbol{\tau_A}$ inertial of surrounding fluid known as added mass, and the restoring force and moment, $\boldsymbol{\tau_G}$. These forces and moments tend to oppose the motion of the ROV and, other than the restoring force, are dependent on the velocities and accelerations of the vehicle and are hence expressed in the body-fixed frame. The restoring force and moment $\boldsymbol{\tau_G}$ due to the weight and buoyancy force have to be transformed to the body-fixed frame forces and moments $\mathbf{G_f}(\boldsymbol{\eta})$. The open-loop nonlinear ROV dynamic equations can be expressed according to:

$$\boldsymbol{\tau} = \boldsymbol{\tau_A} + \boldsymbol{\tau_H} + \mathbf{G_f}(\boldsymbol{\eta}) = \boldsymbol{\tau_A} - \mathbf{M_A} - \mathbf{C_A}(\boldsymbol{\nu}) - \mathbf{D}(\boldsymbol{\nu})\boldsymbol{\nu} - \mathbf{G_f}(\boldsymbol{\eta})$$

$$\mathbf{M}\dot{\boldsymbol{\nu}} + \mathbf{C}(\boldsymbol{\nu})\boldsymbol{\nu} + \mathbf{D}(\boldsymbol{\nu})\boldsymbol{\nu} + \mathbf{G_f}(\boldsymbol{\eta}) = \boldsymbol{\tau_A} \tag{1.15}$$

where $\boldsymbol{\nu} = \begin{bmatrix} u & v & w & p & q & r \end{bmatrix}^{\mathrm{T}}$ is the body-fixed velocity vector and $\boldsymbol{\eta} = \begin{bmatrix} \boldsymbol{\eta_1} & \boldsymbol{\eta_2} \end{bmatrix}^{\mathrm{T}}$ is the earth-fixed vector, comprising the position vector $\boldsymbol{\eta_1} = \begin{bmatrix} x & y & z \end{bmatrix}^{\mathrm{T}} \in \Re^3$ and the orientation vector of Euler angles, $\boldsymbol{\eta_2} = \begin{bmatrix} \phi & \theta & \psi \end{bmatrix}^{\mathrm{T}} \in S^3$ (a torus of dimensions three) in Fig. 1.1; the angles are defined on the interval $[0, 2\pi)$. $\mathbf{M} = \mathbf{M_{RB}} + \mathbf{M_A} \in \Re^{6\times6}$ is the sum of the rigid body inertia mass and added fluid inertia mass matrix, $\mathbf{C}(\boldsymbol{\nu}) = \mathbf{C_{RB}}(\boldsymbol{\nu}) + \mathbf{C_A}(\boldsymbol{\nu}) \in \Re^{6\times6}$ is the sum of Coriolis and centripetal and the added mass forces and moments matrix, and $\mathbf{D}(\boldsymbol{\nu}) \in \Re^{6\times6}$ is the damping matrix due to the surrounding fluid. The input force and moment vector $\boldsymbol{\tau_A} = \mathbf{Tu} \in \Re^6$ relate the thrust output vector $\mathbf{u} = \mathbf{F_T}\bar{\mathbf{u}} \in \Re^4$ with the thruster configuration matrix $\mathbf{T} \in \Re^{6\times4}$, defined in Sect. 1.3; $\mathbf{F_T} \in \Re^{4\times4}$ is the dynamics of each thruster and converts the input voltage command $\bar{\mathbf{u}} \in \Re^4$ into thrusts to propel the vehicle. Here, ROV is assumed to be an underactuated system (i.e. the number of actuators equal to 4 is less than the number of controllable degrees of freedom, i.e. 6).

The following are the assumptions used in the ROV modeling. The ROV is designed to move very slowly and under the assumption of an ideal fluid that non-viscous liquid and the hydrodynamics forces are determined under the assumption of potential flow. In addition, it is not unreasonable to consider only linear and diagonal hydrodynamic damping under these conditions. Main fluid–body interactions are the hydrodynamic forces like added mass and damping. Besides, with an additional balancing mass, the ROV is made to be neutrally buoyant, where the X–Y coordinate of the center of gravity (CG) coincides with the center of buoyancy (CB).

1.1.1 Hydrodynamic Forces: Added Mass Terms

The motion of the surrounding body of fluid in response to the ROV motion manifests itself as hydrodynamic forces and moments that resist the vehicle motion. These forces and moments come about from the acceleration that the fluid particles experience when they encounter the vehicle. As shown below, their effect appears to be like "added" mass and inertia. For a fully submerged vehicle, the added mass and inertia are independent of the wave circular frequency.

Hydrodynamic Added Mass Force X_A acting along the X-axis
The relationship between the hydrodynamic forces and moments and accelerations can be represented by coefficients known as hydrodynamics derivatives. For example, if there is acceleration $\dot{u}$ in the X-direction, the hydrodynamic force X_A arising from that motion can be given as:

$$X_A = X_{\dot{u}}\dot{u} \tag{1.16}$$

where the hydrodynamic derivative $X_{\dot{u}} = \frac{\partial X}{\partial \dot{u}}$. The single DOF equation of motion to describe the ROV moving in the X-direction can be written as: $(m - X_{\dot{u}})\dot{u} = \tau_x + X_u u$; m is the mass, X_u is the linear damping, and τ_x is the thrust. In this form, the hydrodynamic derivative $X_{\dot{u}}$ can be considered as an added mass term on the left-hand side of the expression.

As seen in (1.16), these forces and moments are proportional to the acceleration of the vehicle. In order to consider this effect, the added mass inertia and Coriolis and centripetal matrices that are used in (1.15) are defined. Similar to the rigid-body case, hydrodynamic added mass term also consists of two parts: the added mass inertia component and the added mass Coriolis and centrifugal force component [1]. The hydrodynamic added mass matrix is written as:

$$\mathbf{M_A} = \begin{bmatrix} X_{\dot{u}} & X_{\dot{v}} & X_{\dot{w}} & X_{\dot{p}} & X_{\dot{q}} & X_{\dot{r}} \\ Y_{\dot{u}} & Y_{\dot{v}} & Y_{\dot{w}} & Y_{\dot{p}} & Y_{\dot{q}} & Y_{\dot{r}} \\ Z_{\dot{u}} & Z_{\dot{v}} & Z_{\dot{w}} & Z_{\dot{p}} & Z_{\dot{q}} & Z_{\dot{r}} \\ K_{\dot{u}} & K_{\dot{v}} & K_{\dot{w}} & K_{\dot{p}} & K_{\dot{q}} & K_{\dot{r}} \\ M_{\dot{u}} & M_{\dot{v}} & M_{\dot{w}} & M_{\dot{p}} & M_{\dot{q}} & M_{\dot{r}} \\ N_{\dot{u}} & N_{\dot{v}} & N_{\dot{w}} & N_{\dot{p}} & N_{\dot{q}} & N_{\dot{r}} \end{bmatrix} \tag{1.17}$$

On the other hand, the hydrodynamic added Coriolis and centripetal matrix $\mathbf{C_A}(\boldsymbol{\nu}) = -\mathbf{C_A^T}(\boldsymbol{\nu})$ are given by:

$$\mathbf{C_A}(\boldsymbol{\nu}) = \begin{bmatrix} 0 & 0 & 0 & 0 & -a_3 & a_2 \\ 0 & 0 & 0 & a_3 & 0 & -a_1 \\ 0 & 0 & 0 & -a_2 & a_1 & 0 \\ 0 & -a_3 & a_2 & 0 & -b_3 & b_2 \\ a_3 & 0 & -a_1 & b_3 & 0 & -b_1 \\ -a_2 & a_1 & 0 & -b_2 & b_1 & 0 \end{bmatrix} \tag{1.18}$$

where

$$\begin{aligned} a_1 &= X_{\dot{u}}u + X_{\dot{v}}v + X_{\dot{w}}w + X_{\dot{p}}p + X_{\dot{q}}q + X_{\dot{r}}r \\ a_2 &= X_{\dot{v}}u + Y_{\dot{v}}v + Y_{\dot{w}}w + Y_{\dot{p}}p + Y_{\dot{q}}q + Y_{\dot{r}}r \\ a_3 &= X_{\dot{w}}u - Y_{\dot{w}}v + Z_{\dot{w}}w + Z_{\dot{p}}p + Z_{\dot{q}}q + Z_{\dot{r}}r \\ b_1 &= X_{\dot{p}}u + Y_{\dot{p}}v + Z_{\dot{p}}w + K_{\dot{p}}p + K_{\dot{q}}q + K_{\dot{r}}r \\ b_2 &= X_{\dot{q}}u + Y_{\dot{q}}v + Z_{\dot{q}}w + K_{\dot{q}}p + M_{\dot{q}}q + M_{\dot{r}}r \\ b_3 &= X_{\dot{r}}u + Y_{\dot{r}}v + Z_{\dot{r}}w + K_{\dot{r}}p + M_{\dot{r}}q + N_{\dot{r}}r \end{aligned} \tag{1.19}$$

The hydrodynamic added mass matrix was computed through the use of the following assumptions commonly stated in the literature [1]:

1. To simplify the problem, the ideal fluid is used. It refers to fluid that is incompressible, inviscid (frictionless), and irrotational (fluid particles are not rotating). For a rigid body moving very slowly, the hydrodynamic system inertia matrix, $\mathbf{M_A}$ is positive and constant. The numerical values of the added mass derivatives in a real fluid are usually in good agreement with those obtained from ideal theory. Hence $\mathbf{M_A} > 0$ is a good approximation.
2. The effects of the off-diagonal elements in $\mathbf{M_A}$ on the underwater vehicle are small compared to the diagonal elements. For most low-speed underwater vehicles, the off-diagonal terms are often neglected. This approximation is found to hold true for many applications [1]. Hence, $\mathbf{M_A}$ has the simplified diagonal form as follows:

$$\mathbf{M_A} = -\text{diag}\{X_{\dot{u}}, Y_{\dot{v}}, Z_{\dot{w}}, K_{\dot{p}}, M_{\dot{q}}, N_{\dot{r}}\} \tag{1.20}$$

3. As the off-diagonal elements $\mathbf{M_A}$ are neglected, the Coriolis and centripetal added mass matrix $\mathbf{C_A}(\boldsymbol{\nu})$ is simplified to:

$$\mathbf{C_A}(\boldsymbol{\nu}) = \begin{bmatrix} 0 & 0 & 0 & 0 & -Z_{\dot{w}}w & Y_{\dot{v}}v \\ 0 & 0 & 0 & Z_{\dot{w}}w & 0 & -X_{\dot{u}}u \\ 0 & 0 & 0 & -Y_{\dot{v}}v & X_{\dot{u}}u & 0 \\ 0 & -Z_{\dot{w}}w & Y_{\dot{v}}v & 0 & -N_{\dot{r}}r & M_{\dot{q}}q \\ Z_{\dot{w}}w & 0 & -X_{\dot{u}}u & N_{\dot{r}}r & 0 & -K_{\dot{p}}p \\ -Y_{\dot{v}}v & X_{\dot{u}}u & 0 & -M_{\dot{q}}q & K_{\dot{p}}p & 0 \end{bmatrix} \tag{1.21}$$

1.1.2 Hydrodynamic Damping Matrix

The matrix $\mathbf{D}(\boldsymbol{\nu})$ is the hydrodynamic damping matrix consisting of both linear and quadratic terms. This hydrodynamic damping is caused by potential damping due to skin friction (linear damping) drag and vortex shedding (quadratic damping). Sum of these individual components gives the overall hydrodynamic damping effect on the ROV.

$$\mathbf{D_Q} = -\text{diag}\{X_{u|u|}, Y_{v|v|}, Z_{w|w|}, K_{p|p|}, M_{q|q|}, N_{r|r|}\} \tag{1.22a}$$

The hydrodynamic damping matrix can be simplified as follows:

1. As the ROV is operating at around the linear speed of 0.2 m/s (or angular speed of 0.2 rad/s), it is well within the linear damping region of maximum 2 m/s. Hence, the linear hydrodynamic damping is used.
2. The off-diagonal elements in **D** on an underwater vehicle are small compared to the diagonal elements, hence only the hydrodynamic damping in the diagonal form is used:

$$\mathbf{D_L} = -\text{diag}\{X_u, Y_v, Z_w, K_p, M_q, N_r\} \tag{1.22b}$$

1.2 Kinematics Equation and Earth-Fixed Frame Equation Using Euler's Angles

The transformation using the Euler's angles provides an important transformation between the dynamics expressed in the body-fixed frame in (1.15) to the earth-fixed frame as seen in Fig. 1.1. As the accelerations of a point on the surface of the earth can be neglected, for slow-moving marine vehicles, the earth-fixed frame can be considered to be an inertial frame. The kinematics equations [1] that represent the Euler's transformation can be written as:

$$\dot{\boldsymbol{\eta}} = \mathbf{J}(\boldsymbol{\eta})\boldsymbol{\nu} \tag{1.23}$$

Alternatively, the Euler's transformation between the body and earth-fixed frames $(\boldsymbol{\eta}, \boldsymbol{\nu})$ to $(\boldsymbol{\eta}, \dot{\boldsymbol{\eta}})$ can be written as:

$$\begin{bmatrix} \boldsymbol{\eta} \\ \dot{\boldsymbol{\eta}} \end{bmatrix} = \begin{bmatrix} \mathbf{I}_{6\times6} & 0_{6\times6} \\ 0_{6\times6} & \mathbf{J}(\boldsymbol{\eta}_2) \end{bmatrix} \begin{bmatrix} \boldsymbol{\eta} \\ \boldsymbol{\nu} \end{bmatrix} \tag{1.24}$$

where the Euler transformation matrix, $\mathbf{J}(\boldsymbol{\eta})$ is derived by successive rotation of the Euler's angles ($\boldsymbol{\eta}_2 = \begin{bmatrix} \phi & \theta & \psi \end{bmatrix}^{\mathrm{T}}$) about Z, Y, and X axes in Fig. 1.1 to give:

$$\mathbf{J}(\boldsymbol{\eta}) = \begin{bmatrix} \mathbf{J}_1(\boldsymbol{\eta}_2) & 0 \\ 0 & \mathbf{J}_2(\boldsymbol{\eta}_2) \end{bmatrix} \tag{1.25}$$

where

$$\mathbf{J}_1(\boldsymbol{\eta}_2) = \begin{bmatrix} \mathrm{c}(\psi)\mathrm{c}(\theta) & -\mathrm{s}(\psi)\mathrm{c}(\phi) + \mathrm{c}(\psi)\mathrm{s}(\theta)\mathrm{s}(\phi) & \mathrm{s}(\psi)\mathrm{s}(\phi) + \mathrm{c}(\psi)\mathrm{c}(\phi)\mathrm{s}(\theta) \\ \mathrm{s}(\psi)\mathrm{c}(\theta) & \mathrm{c}(\psi)\mathrm{c}(\phi) + \mathrm{s}(\phi)\mathrm{s}(\theta)\mathrm{s}(\psi) & -\mathrm{c}(\psi)\mathrm{s}(\phi) + \mathrm{s}(\theta)\mathrm{s}(\psi)\mathrm{c}(\theta) \\ -\mathrm{s}(\theta) & \mathrm{c}(\theta)\mathrm{s}(\phi) & \mathrm{c}(\theta)\mathrm{c}(\phi) \end{bmatrix};$$

$$\mathbf{J}_2(\boldsymbol{\eta}_2) = \begin{bmatrix} 1 & \mathrm{s}(\phi)\mathrm{t}(\theta) & \mathrm{c}(\phi)\mathrm{t}(\theta) \\ 0 & \mathrm{c}(\phi) & -\mathrm{s}(\phi) \\ 0 & \frac{\mathrm{s}(\phi)}{\mathrm{c}(\theta)} & \frac{\mathrm{c}(\phi)}{\mathrm{c}(\theta)} \end{bmatrix}$$

and $\mathrm{s} = \sin(\cdot)$, $\mathrm{c} = \cos(\cdot)$, and $\mathrm{t} = \tan(\cdot)$. The transformation is undefined for $\theta = \pm 90°$. To overcome this singularity, a quaternion approach must be considered. However, for the ROV, this problem does not exist because the vehicle is neither designed nor required to operate at $\theta = \pm 90°$.

Differentiating (1.24) and rearranging the equation gives:

$$\begin{bmatrix} \boldsymbol{\nu} \\ \ddot{\eta} \end{bmatrix} = \begin{bmatrix} \mathbf{J}^{-1}(\boldsymbol{\eta}_2) & \mathbf{0}_{6\times 6} \\ \dot{\mathbf{J}}(\boldsymbol{\eta}_2)\mathbf{J}^{-1}(\boldsymbol{\eta}_2) & \mathbf{J}(\boldsymbol{\eta}_2) \end{bmatrix} \begin{bmatrix} \dot{\boldsymbol{\eta}} \\ \dot{\boldsymbol{\nu}} \end{bmatrix} \tag{1.26}$$

where the body-fixed dynamic can be transformed into the earth-fixed frame (or vice versa) by substituting the $\boldsymbol{\nu}$ and $\dot{\boldsymbol{\nu}}$ as:

$$\boldsymbol{\nu} = \mathbf{J}^{-1}(\boldsymbol{\eta})\dot{\boldsymbol{\eta}} \tag{1.27}$$

$$\begin{aligned} \dot{\boldsymbol{\nu}} &= \mathbf{J}^{-1}(\boldsymbol{\eta})\ddot{\boldsymbol{\eta}} - \mathbf{J}^{-2}(\boldsymbol{\eta})\dot{\mathbf{J}}(\boldsymbol{\eta})\dot{\boldsymbol{\eta}} \\ &= \mathbf{J}^{-1}(\boldsymbol{\eta})[\ddot{\boldsymbol{\eta}} - \dot{\mathbf{J}}(\boldsymbol{\eta})\mathbf{J}^{-1}(\boldsymbol{\eta})\dot{\boldsymbol{\eta}}] \end{aligned} \tag{1.28}$$

into the body-fixed dynamic gives the ROV's dynamic in the earth-fixed frame:

$$\mathbf{M}_{\eta}(\boldsymbol{\eta})\ddot{\boldsymbol{\eta}} + \mathbf{C}_{\eta}(\boldsymbol{\nu}, \boldsymbol{\eta})\dot{\boldsymbol{\eta}} + \mathbf{D}_{\eta}(\boldsymbol{\nu}, \boldsymbol{\eta})\dot{\boldsymbol{\eta}} + \mathbf{G}_{\eta}(\boldsymbol{\eta}) = \boldsymbol{\tau}_{\eta} \tag{1.29}$$

where

$$\mathbf{M}_{\eta}(\boldsymbol{\eta}) = \mathbf{J}^{-\mathrm{T}}(\boldsymbol{\eta})\mathbf{M}\mathbf{J}^{-1}(\boldsymbol{\eta}),$$

$$\mathbf{C}_\eta(\mathbf{v}, \eta) = \mathbf{J}^{-T}(\eta)[\mathbf{C}(\mathbf{v}) - \mathbf{M}\mathbf{J}^{-1}(\eta)\dot{\mathbf{J}}(\eta)]\mathbf{J}^{-1}(\eta)$$

$$\mathbf{D}_\eta(\mathbf{v}, \eta) = \mathbf{J}(\eta)^{-T}\mathbf{D}\mathbf{J}(\eta)^{-1}$$

$$\mathbf{G}_\eta(\eta) = \mathbf{J}^{-T}(\eta)\mathbf{G}_\mathbf{f}(\eta)$$

$$\tau_\eta = \mathbf{J}^{-T}(\eta)\tau_\mathbf{A}.$$

Hence, the dynamic in the earth-fixed frame is derived as seen in (1.29).

1.2.1 Buoyancy and Gravitational Force Vector

The term $\mathbf{G}_\mathbf{f}(\eta)$ is used to describe the gravitational and buoyancy [1] vector exerted on the ROV in water. Transforming the restoring forces to the body-fixed coordinate system, the weight and buoyancy forces can be written as:

$$\mathbf{f}_\mathbf{G}(\eta) = \mathbf{J}_1^T(\eta)\begin{bmatrix} 0 \\ 0 \\ W \end{bmatrix} \tag{1.30}$$

$$\mathbf{f}_\mathbf{B}(\eta) = -\mathbf{J}_1^T(\eta)\begin{bmatrix} 0 \\ 0 \\ B \end{bmatrix} \tag{1.31}$$

where the weight of the ROV is given by $W = mg$ and g is the gravitational acceleration (positive downward) term. When fully submerged, the ROV's buoyancy is equal to the weight of water displaced, i.e., $B = \rho g \nabla$, where ρ is the fluid density and ∇ is the volume displaced by the submerged ROV. Here, $\mathbf{J}_1(\eta)$ is the Euler angle coordinate transformation matrix. In the body-fixed coordinate system, the restoring force vector becomes:

$$\mathbf{G}_\mathbf{f}(\eta) = -\begin{bmatrix} \mathbf{f}_\mathbf{G}(\eta) + \mathbf{f}_\mathbf{B}(\eta) \\ \mathbf{r}_\mathbf{G} \times \mathbf{f}_\mathbf{G}(\eta) + \mathbf{r}_\mathbf{B} \times \mathbf{f}_\mathbf{B}(\eta) \end{bmatrix} \tag{1.32}$$

Expanding the preceding equation gives:

$$\mathbf{G_f}(\eta) = \begin{bmatrix} (W - B)\sin\theta \\ -(W - B)\cos\theta\sin\phi \\ -(W - B)\cos\theta\cos\phi \\ -(y_G W - y_B B)\cos\theta\cos\phi + (z_G W - z_B B)\cos\theta\sin\phi \\ (z_G W - z_B B)\sin\theta + (x_G W - x_B B)\cos\theta\cos\phi \\ -(x_G W - x_B B)\cos\theta\sin\phi - (y_G W - y_B B)\sin\theta \end{bmatrix} \tag{1.33}$$

The buoyancy and gravitational force hydrodynamic damping matrix, $\mathbf{G_f}(\eta)$ was computed using the following design rules:

1. In designing ROVs, it is desirable to make the ROV neutrally buoyant or slightly positive buoyant by adding additional float or balancing mass. With that, the ROV becomes neutrally buoyant, W = B.
2. By placing additional mass on the ROV to make the X–Y coordinates of the C.B. coincide with the X–Y coordinate of the CG, that is, $x_G = x_B = 0$, $y_G = y_B = 0$. The resulting equation becomes:

$$\mathbf{G_f}(\eta) = \begin{bmatrix} 0 \\ 0 \\ 0 \\ (z_G - z_B)W\cos\theta\sin\phi \\ (z_G - z_B)W\sin\theta \\ 0 \end{bmatrix} \tag{1.34}$$

1.3 Thruster Configuration Matrix

The position of the thruster on the ROV can be defined by the thrusters' configuration matrix. This is followed by establishing the thrust versus the input voltage relationship model. As mentioned in the nonlinear ROV dynamic equation, the left-hand side of the equations refers to the input forces and moments to the ROV. These input forces and moments are determined based on the summation of the force and moment equations in the six DOFs.

For example, the input forces (along X, Y, and Z axes) and moments (about X, Y, and Z axes) from the four thrusters can be written as follows:

$$\begin{aligned} u_1 + u_2 &= \tau_x \\ u_3\sin\beta - u_4\sin\beta &= \tau_y \\ u_3\cos\beta + u_4\cos\beta &= \tau_z \\ u_3\delta - u_4\delta &= \tau_\phi \end{aligned}$$

$$
\begin{aligned}
-\mathrm{u}_1\varepsilon - \mathrm{u}_2\varepsilon + \mathrm{u}_3\alpha\cos\beta - \mathrm{u}_4\alpha\cos\beta &= \tau_\theta \\
\mathrm{u}_1\gamma - \mathrm{u}_2\gamma + \mathrm{u}_3\alpha\sin\beta - \mathrm{u}_4\alpha\sin\beta &= \tau_\psi
\end{aligned} \tag{1.35}
$$

where α, β, γ, δ, and ε can be geometrical parameters based on the thrusters' location on the ROV platform. The thrusters' inputs such as $\mathrm{u}_{\min} \le \mathrm{u}_i \le \mathrm{u}_{\max}, i = 1, 2, 3, 4$ are the minimum and maximum thrust output by each thruster and $\boldsymbol{\tau}_\mathbf{A} = [\tau_x\ \tau_y\ \tau_z\ \tau_\phi\ \tau_\theta\ \tau_\psi]^\mathrm{T}$ is the force and moment vector generated in the six DOFs.

To simplify (1.35) into matrix form:

$$
\begin{bmatrix} \tau_x \\ \tau_y \\ \tau_z \\ \tau_\phi \\ \tau_\theta \\ \tau_\psi \end{bmatrix} = \overbrace{\begin{bmatrix} 1 & 1 & 0 & 0 \\ 0 & 0 & \sin\beta & -\sin\beta \\ 0 & 0 & \cos\beta & \cos\beta \\ 0 & 0 & \delta & -\delta \\ -\varepsilon & -\varepsilon & \alpha\cos\beta & -\alpha\cos\beta \\ \gamma & -\gamma & \alpha\sin\beta & -\alpha\sin\beta \end{bmatrix}}^{\mathbf{T}} \overbrace{\begin{bmatrix} \mathrm{u}_1 \\ \mathrm{u}_2 \\ \mathrm{u}_3 \\ \mathrm{u}_4 \end{bmatrix}}^{\mathbf{u}}
$$

$$
\boldsymbol{\tau}_\mathbf{A} = \mathbf{Tu} \tag{1.36}
$$

where $\mathbf{u} = \mathbf{F_T}\bar{\mathbf{u}} \in \Re^4$ with $\mathbf{F_T} = f_T\mathbf{I}_4, \bar{\mathbf{u}} \in \Re^4$ is the commanded voltage input to the thrusters, and $\mathbf{I}_4$ is an identity matrix of dimension four. In summary, the relationship between these parameters can be shown in the block diagram in Fig. 1.2.

A steady-state thruster model can be considered due to its faster response time as compared to the ROV dynamic. An experiment to establish the thruster's steady-state model (f_T) between the thrust ($\mathbf{u}$) and the commanded voltage input ($\bar{\mathbf{u}}$) can be considered. Sometimes, the thruster has a relatively shorter response time compared with the ROV and thus can be modeled as a gain term. In some cases, it is quite negligible. Hence, the thruster's steady-state model can be considered as $f_T = 1$.

As mentioned, the thrusters can be less than the number of controllable degrees of freedom. The following least-square optimization method [2] is used. Assume that the vector $\boldsymbol{\tau}_\mathbf{A}$ is bounded in such a way that the calculated elements of the vector $\mathbf{u}$ can never exceed the boundary values $\mathbf{u}_{\min}$ and $\mathbf{u}_{\max}$. Then the unconstrained thrust

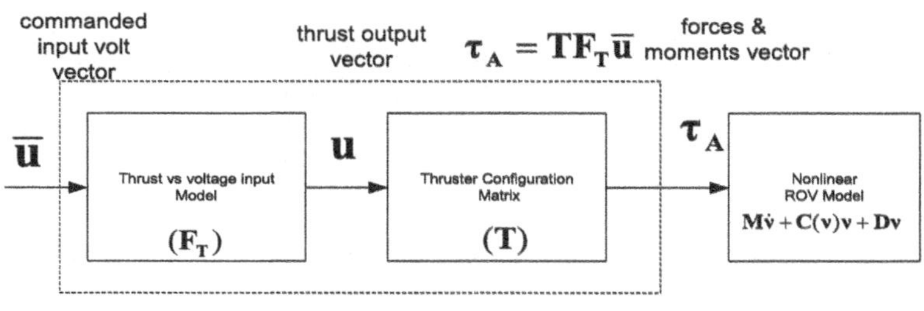

Fig. 1.2 Block diagram of thrusters' parameters

allocation problem can be formulated as the following least-squares optimization problem:

$$\min_{\mathbf{u}} = \frac{1}{2}\mathbf{u}^{\mathrm{T}}\mathbf{W}\mathbf{u} \tag{1.37}$$

subject to

$$\boldsymbol{\tau}_{\mathbf{A}} - \mathbf{T}\mathbf{u} = 0 \tag{1.38}$$

where $\mathbf{W}$ is a positive definite matrix.

The solution of the above problem using Lagrange multipliers is given by

$$\mathbf{u} = \mathbf{T}^{*}\boldsymbol{\tau}_{\mathbf{A}} \tag{1.39}$$

where

$$\mathbf{T}^{*} = \mathbf{W}^{-1}\mathbf{T}^{\mathrm{T}}(\mathbf{T}\mathbf{W}^{-1}\mathbf{T}^{\mathrm{T}})^{-1} \tag{1.40}$$

is recognized as the generalized inverse. For the case $\mathbf{W} = \mathbf{I}$, that is equally weighted control forces, the expression (1.40) reduces to the Moore–Penrose pseudoinverse:

$$\mathbf{T}^{*} = \mathbf{T}^{\mathrm{T}}(\mathbf{T}\mathbf{T}^{\mathrm{T}})^{-1} \tag{1.41}$$

Since $\mathbf{u} = \mathbf{T}^{*}\boldsymbol{\tau}_{\mathbf{A}}$, the control input vector $\mathbf{u}$ can be computed as:

$$\mathbf{u} = \mathbf{T}^{*}\boldsymbol{\tau}_{\mathbf{A}} \tag{1.42}$$

1.4 Nonlinear ROV Model

The ROV dynamics and kinematics given in Sects. 1.1 and 1.2 are:

$$\mathbf{M}\dot{\boldsymbol{\nu}} + \mathbf{C}(\boldsymbol{\nu}) + \mathbf{D}(\boldsymbol{\nu})\boldsymbol{\nu} + \mathbf{G}_{\mathbf{f}}(\boldsymbol{\eta}) = \boldsymbol{\tau}_{\mathbf{A}} \tag{1.43}$$

$$\dot{\boldsymbol{\eta}} = \mathbf{J}(\boldsymbol{\eta})\boldsymbol{\nu} \tag{1.44}$$

These will be expressed in the state-space form to facilitate the analysis and design of model-based controllers.

$$\dot{\mathbf{x}} = \mathbf{f}(\mathbf{x}, t) + \mathbf{g}(\bar{\mathbf{u}}, t), \mathbf{x}(0) = \mathbf{x}_0 \tag{1.45}$$

The full-order states can be written as follows:

$$\mathbf{x} = \begin{bmatrix} \boldsymbol{\eta} & \boldsymbol{\nu} \end{bmatrix}^{\mathrm{T}}$$

$$\mathbf{f}(\mathbf{x}, t) = \begin{bmatrix} \mathbf{0}_{6\times6} & \mathbf{J}(\boldsymbol{\eta}) \\ \mathbf{0}_{6\times6} & -\mathbf{M}^{-1}[\mathbf{C}(\boldsymbol{\nu}) + \mathbf{D}(\boldsymbol{\nu})] \end{bmatrix} \begin{bmatrix} \boldsymbol{\eta} \\ \boldsymbol{\nu} \end{bmatrix} + \begin{bmatrix} \mathbf{0}_{6\times1} \\ -\mathbf{M}^{-1}\mathbf{G}_{\mathbf{f}}(\boldsymbol{\eta}) \end{bmatrix}$$

$$\mathbf{g}(\bar{\mathbf{u}}, t) = \begin{bmatrix} \mathbf{0}_{6\times1} \\ \mathbf{M}^{-1}\mathbf{T}\mathbf{F}_{\mathbf{T}}\bar{\mathbf{u}} \end{bmatrix}$$

The output equation can be expressed as:

$$\mathbf{y} = \mathbf{h}(\mathbf{x}, t) \tag{1.46}$$

where

$$\mathbf{h}(\mathbf{x}, t) = \begin{bmatrix} \mathbf{I}_{6\times6} & \mathbf{0}_{6\times6} \\ \mathbf{0}_{6\times6} & \mathbf{0}_{6\times6} \end{bmatrix} \mathbf{x}$$

The state vector $\mathbf{f}(\mathbf{x}, t)$ is piecewise continuous in t and locally Lipschitz in $\mathbf{x}$, $\mathbf{g}(\bar{\mathbf{u}}, t)$ is piecewise continuous in t and locally Lipschitz in $\bar{\mathbf{u}}$, $\mathbf{h}(\mathbf{x}, t)$ is piecewise continuous in t and continuous in $\mathbf{x}$.

The parameters used in $\mathbf{f}(\mathbf{x}, t)$ can be determined. They are observed to have the following properties and are consistent with [1].

The matrices in (1.29) and (1.30) satisfy the following property:

(a) $\mathbf{M} = \mathbf{M}^{\mathrm{T}} > 0, \dot{\mathbf{M}} = 0$ where $\mathbf{M} = \mathbf{M}_{\mathbf{RB}} + \mathbf{M}_{\mathbf{A}}$
(b) $\mathbf{C}(\boldsymbol{\nu}) = -\mathbf{C}^{\mathrm{T}}(\boldsymbol{\nu}), \boldsymbol{\nu} \in \Re^6$ where $\mathbf{C}(\boldsymbol{\nu}) = \mathbf{C}_{\mathbf{RB}}(\boldsymbol{\nu}) + \mathbf{C}_{\mathbf{A}}(\boldsymbol{\nu})$

(c) $s^{\mathrm{T}}[\dot{\mathbf{M}} - 2\mathbf{C}(\boldsymbol{\nu})]s = 0, \forall s \in \Re^6$

(d) $\mathbf{D}(\boldsymbol{\nu}) > 0$

1.5 Station-Keeping Model

A station-keeping model is used to describe the ROV dynamic during the hovering or commonly known as stabilization about equilibrium. As the roll and pitch motions are unspecified in the equilibrium and they are asymptotically stable and bounded, they are neglected in the station-keeping mode. It mainly involves four DOFs such as surge, sway, heave, and yaw. In this section, the nominal and perturbed models for the station-keeping dynamics are formulated.

The nominal nonlinear dynamic equation of the station-keeping model can be written as:

$$(m - X_{\dot{u}})\dot{u} + X_u u + (Y_{\dot{v}} - m)vr = \tau_x \tag{1.47}$$

$$(m - Y_{\dot{v}})\dot{v} + Y_v v + (m - X_{\dot{u}})ur = \tau_y \tag{1.48}$$

$$(m - Z_{\dot{w}})\dot{w} + Z_w w = \tau_z \tag{1.49}$$

$$(I_z - N_{\dot{r}})\dot{r} + N_r r + (X_{\dot{u}}u - Y_{\dot{v}}u)v = \tau_\psi \tag{1.50}$$

The kinematics equation can be written as:

$$\dot{x} = u\cos\psi - v\sin\psi \tag{1.51}$$

$$\dot{y} = u\sin\psi + v\cos\psi \tag{1.52}$$

$$\dot{z} = w \tag{1.53}$$

$$\dot{\psi} = r \tag{1.54}$$

In compact form, the nonlinear dynamic equation in (1.47)–(1.50) can be written as (variables labeled with subscript "s"):

$$\dot{\mathbf{v}}_\mathrm{s} = -\mathbf{M}_\mathrm{s}^{-1}[\mathbf{C}_\mathrm{s}(\mathbf{v}_\mathrm{s}) + \mathbf{D}_\mathrm{s}]\mathbf{v}_\mathrm{s} + \mathbf{M}_\mathrm{s}^{-1}\boldsymbol{\tau}_\mathrm{s} \tag{1.55}$$

where $\mathbf{v}_\mathrm{s} = \begin{bmatrix} u & v & w & r \end{bmatrix}^\mathrm{T}$, $\boldsymbol{\tau}_\mathrm{s} = \mathbf{T}_\mathrm{s}\mathbf{F}_\mathbf{T}\bar{\mathbf{u}} = [\tau_x\ \tau_y\ \tau_z\ \tau_\psi]^\mathrm{T}$ in which $\bar{\mathbf{u}} \in \Re^4$, $\mathbf{F}_\mathbf{T} \in \Re^{4\times4}$,

$$\mathbf{T}_\mathrm{s} = \begin{bmatrix} 1 & 1 & 0 & 0 \\ 0 & 0 & \sin\beta & -\sin\beta \\ 0 & 0 & \cos\beta & \cos\beta \\ \gamma & -\gamma & \alpha\sin\beta & -\alpha\sin\beta \end{bmatrix}; \quad \mathbf{M}_\mathrm{s} = \begin{bmatrix} m - X_{\dot{u}} & 0 & 0 & 0 \\ 0 & m - Y_{\dot{v}} & 0 & 0 \\ 0 & 0 & m - Z_{\dot{w}} & 0 \\ 0 & 0 & 0 & I_r - N_{\dot{r}} \end{bmatrix};$$

$$\mathbf{D}_\mathrm{s} = \begin{bmatrix} X_u & 0 & 0 & 0 \\ 0 & Y_v & 0 & 0 \\ 0 & 0 & Z_w & 0 \\ 0 & 0 & 0 & N_r \end{bmatrix}; \quad \mathbf{C}_\mathrm{s}(\mathbf{v}_\mathrm{s}) = \begin{bmatrix} 0 & (Y_{\dot{v}} - m)r & 0 & 0 \\ 0 & 0 & 0 & (m - X_{\dot{u}})u \\ 0 & 0 & 0 & 0 \\ 0 & -Y_{\dot{v}}u + X_{\dot{u}}u & 0 & 0 \end{bmatrix}$$

The kinematics equation in (1.51)–(1.54) can be written in compact form as:

$$\dot{\boldsymbol{\eta}}_s = \mathbf{J}_s(\boldsymbol{\eta}_s)\boldsymbol{\nu}_s \tag{1.56}$$

where $\boldsymbol{\eta}_s = \begin{bmatrix} x & y & z & \psi \end{bmatrix}^T$ and $\mathbf{J}_s(\boldsymbol{\eta}_s) = \begin{bmatrix} \cos\psi & -\sin\psi & 0 & 0 \\ \sin\psi & \cos\psi & 0 & 0 \\ 0 & 0 & 1 & 0 \\ 0 & 0 & 0 & 1 \end{bmatrix}$.

Equations (1.55) and (1.56) can be rewritten in state-space form as:

$$\dot{\mathbf{x}}_s = \bar{\mathbf{f}}_s(\mathbf{x}_s, t) + \bar{\mathbf{g}}_s(\bar{\mathbf{u}}, t) \tag{1.57}$$

where $\mathbf{x}_s = [\boldsymbol{\eta}_s \; \boldsymbol{\nu}_s]^T$ and

$$\bar{\mathbf{f}}_s(\mathbf{x}_s, t) = \begin{bmatrix} \mathbf{J}_s(\boldsymbol{\eta}_s)\mathbf{v}_s \\ -\mathbf{M}_s^{-1}[\mathbf{C}_s(\boldsymbol{\nu}_s) + \mathbf{D}_s]\boldsymbol{\nu}_s \end{bmatrix} \tag{1.58}$$

and

$$\bar{\mathbf{g}}_s(\bar{\mathbf{u}}, t) = \begin{bmatrix} \mathbf{0}_{3\times 1} \\ \begin{bmatrix} \bar{\mathbf{g}}_{s1} & \bar{\mathbf{g}}_{s2} & \bar{\mathbf{g}}_{s3} & \bar{\mathbf{g}}_{s4} \end{bmatrix}\mathbf{T}_s\mathbf{F}_T\bar{\mathbf{u}} \end{bmatrix} \tag{1.59}$$

where

$$\bar{\mathbf{g}}_{s1} = \begin{bmatrix} \frac{1}{m - X_{\dot{u}}} & 0 & 0 & 0 \end{bmatrix}^T; \quad \bar{\mathbf{g}}_{s2} = \begin{bmatrix} 0 & \frac{1}{m - Y_{\dot{v}}} & 0 & 0 \end{bmatrix}^T; \tag{1.60}$$

$$\bar{\mathbf{g}}_{s3} = \begin{bmatrix} 0 & 0 & \frac{1}{m - Z_{\dot{w}}} & 0 \end{bmatrix}^T; \quad \bar{\mathbf{g}}_{s4} = \begin{bmatrix} 0 & 0 & 0 & \frac{1}{I_r - N_{\dot{r}}} \end{bmatrix}^T \tag{1.61}$$

1.6 ROV Models in Different Planes

From the ROV mass inertia, centripetal and Coriolis matrices, it can be seen that the nonlinear ROV motions are highly coupled, although the roll and pitch motions are designed to be self-stabilizable. To further reduce the motion coupling, it is suggested to adopt a decoupled (see Fig. 1.3) maneuver scheme: moving with a horizontal plane motion whereby the vehicle is first steered at constant forward speed along a straight line that passes through the desired target points at a certain vehicle orientation. At the end of this horizontal plane motion, the vehicle is switched to a vertical plane motion: it maintains a constant orientation (heading) and follows a straight line with a change in altitude. Thus considering only few degrees of freedoms at a time decouples the motions. This reduces the number of controllable degree of freedoms involved in

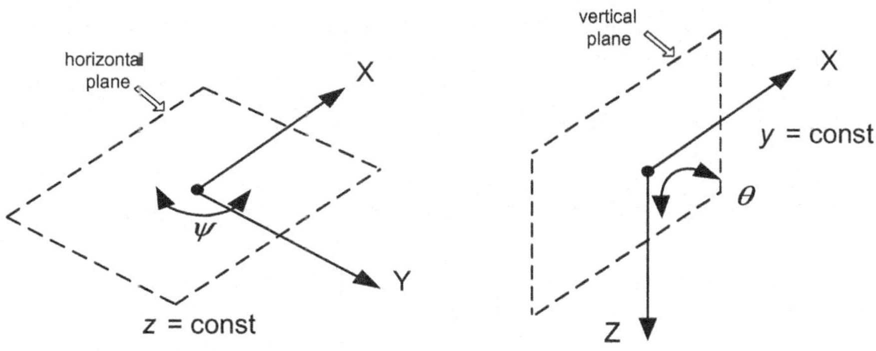

Fig. 1.3 Horizontal (left) and vertical plane (right) motion

subsequent control system analysis and design. With that, the subsystems become actuated.

In summary, the ROV motions can be decoupled into two simple subsystems as follows:

- Horizontal plane subsystem: velocity states (u, v, r) and position states (x, y, ψ).
- Vertical plane subsystem: velocity states (u, w, q) and position states (x, z, θ).

The horizontal plane model describes the ROV dynamic during longitudinal motion that encompasses three DOFs such as: surge, sway, and yaw. The roll, pitch, and heave velocities are bounded as the ROV along the X-axis with z as constant (i.e. $w = \dot{w} = 0$) and at a constant surge velocity that defines the horizontal plane dynamic. With the origin of the ROV coinciding with the center of gravity, the nominal horizontal plane dynamic can be written as:

$$(m - X_{\dot{u}})\dot{u} + X_u u + (Y_{\dot{v}} - m)vr = \tau_x \tag{1.62}$$

$$(m - Y_{\dot{v}})\dot{v} + Y_v v + (m - X_{\dot{u}})ur = \tau_y \tag{1.63}$$

$$(I_z - N_{\dot{r}})\dot{r} + N_r r + (X_{\dot{u}}u - Y_{\dot{v}}u)v = \tau_{\psi} \tag{1.64}$$

The dynamics in u and v are coupled with r, respectively. The kinematics equation becomes:

$$\dot{x} = u\cos\psi - v\sin\psi \tag{1.65}$$

$$\dot{y} = u\sin\psi + v\cos\psi \tag{1.66}$$

$$\dot{\psi} = r \tag{1.67}$$

The dynamics of the horizontal plane (denoted by subscript "h") subsystem as seen in (1.62)–(1.64) can be written in compact form as:

$$\mathbf{M_h}\dot{\boldsymbol{\nu}}_\mathbf{h} + \mathbf{C_h}(\boldsymbol{\nu}_\mathbf{h})\boldsymbol{\nu}_\mathbf{h} + \mathbf{D_h}\boldsymbol{\nu}_\mathbf{h} = \boldsymbol{\tau}_\mathbf{h} \tag{1.68}$$

where $\boldsymbol{\eta}_\mathbf{h} = [x\ y\ \psi]^\mathrm{T}$, $\boldsymbol{\nu}_\mathbf{h} = \begin{bmatrix} u\ v\ r \end{bmatrix}^\mathrm{T}$, $\boldsymbol{\tau}_\mathbf{h} = \mathbf{T_h F_T \bar{u}} = [\,\tau_x\ \tau_y\ \tau_\psi\,]^\mathrm{T}$ and

$$\mathbf{M_h} = \begin{bmatrix} m - X_{\dot{u}} & 0 & 0 \\ 0 & m - Y_{\dot{v}} & 0 \\ 0 & 0 & I_r - N_{\dot{r}} \end{bmatrix}; \quad \mathbf{C_h}(\boldsymbol{\nu}_\mathbf{h}) = \begin{bmatrix} 0 & (Y_{\dot{v}} - m)r & 0 \\ 0 & 0 & (m - X_{\dot{u}})u \\ 0 & -(Y_{\dot{v}} - X_{\dot{u}})u & 0 \end{bmatrix}$$

$$\mathbf{D_h} = \begin{bmatrix} X_u & 0 & 0 \\ 0 & Y_v & 0 \\ 0 & 0 & N_r \end{bmatrix}; \quad \mathbf{T_h} = \begin{bmatrix} 1 & 1 & 0 & 0 \\ 0 & 0 & \sin\beta & -\sin\beta \\ \gamma & -\gamma & \alpha\sin\beta & -\alpha\sin\beta \end{bmatrix}$$

The kinematics equations of motion in the horizontal plane (1.62)–(1.64) can be written as:

$$\dot{\boldsymbol{\eta}}_\mathbf{h} = \mathbf{J_h}(\boldsymbol{\eta}_\mathbf{h})\boldsymbol{\nu}_\mathbf{h} \tag{1.69}$$

where

$$\mathbf{J_h}(\boldsymbol{\eta}_\mathbf{h}) = \begin{bmatrix} \cos\psi & -\sin\psi & 0 \\ \sin\psi & \cos\psi & 0 \\ 0 & 0 & 1 \end{bmatrix}$$

Equations (1.68) and (1.69) can be rewritten in compact form as:

$$\dot{\mathbf{x}}_\mathbf{h} = \bar{\mathbf{f}}_\mathbf{h}(\mathbf{x_h}, t) + \bar{\mathbf{g}}_\mathbf{h}(\bar{\mathbf{u}}, t) \tag{1.70}$$

where $\mathbf{x_h} = [\boldsymbol{\eta}_\mathbf{h}\ \boldsymbol{\nu}_\mathbf{h}]^\mathrm{T}$

$$\bar{\mathbf{f}}_\mathbf{h}(\mathbf{x_h}, t) = \begin{bmatrix} \mathbf{J_h}(\boldsymbol{\eta}_\mathbf{h})\boldsymbol{\nu}_\mathbf{h} \\ -\mathbf{M_h^{-1}}[\mathbf{C_h}(\boldsymbol{\nu}_\mathbf{h}) + \mathbf{D_h}]\boldsymbol{\nu}_\mathbf{h} \end{bmatrix}; \quad \bar{\mathbf{g}}_\mathbf{h}(\bar{\mathbf{u}}, t) = \begin{bmatrix} \mathbf{0}_{3\times1} \\ [\,\bar{\mathbf{g}}_\mathbf{h1}\ \bar{\mathbf{g}}_\mathbf{h2}\ \bar{\mathbf{g}}_\mathbf{h3}\,]\mathbf{T_h F_T \bar{u}} \end{bmatrix};$$

$$\bar{\mathbf{g}}_\mathbf{h1} = \begin{bmatrix} \frac{1}{m - X_{\dot{u}}} & 0 & 0 \end{bmatrix}^\mathrm{T}; \quad \bar{\mathbf{g}}_\mathbf{h2} = \begin{bmatrix} 0 & \frac{1}{m - Y_{\dot{v}}} & 0 \end{bmatrix}^\mathrm{T}; \quad \bar{\mathbf{g}}_\mathbf{h3} = \begin{bmatrix} 0 & 0 & \frac{1}{I_r - N_{\dot{r}}} \end{bmatrix}^\mathrm{T}$$

On the other hand, the vertical plane model describes the ROV dynamic during altitudinal motion that involves three DOFs such as: surge, heave, and pitch. Now, the roll, pitch, and yaw velocities are bounded as the ROV is moving along the Z-axis with a constant y (i.e. $v = \dot{v} = 0$) defining the vertical plane motion dynamic. The nominal vertical plane dynamic for both the translational and rotational direction can be written as:

$$(m - X_{\dot{u}})\dot{u} + X_u u + (m - Z_{\dot{w}})wq = \tau_x \tag{1.71}$$

$$(m - Z_{\dot{w}})\dot{w} + Z_w w + (X_{\dot{u}} - m)uq = \tau_z \tag{1.72}$$

$$(I_y - M_{\dot{q}})\dot{q} + M_q q + (Z_{\dot{w}} - X_{\dot{u}})uw + (z_G - z_B)W\sin\theta = \tau_\theta \tag{1.73}$$

The kinematics equation becomes:

$$\dot{x} = u\cos\theta + w\sin\theta \tag{1.74}$$

$$\dot{z} = -u\sin\theta + w\cos\theta \tag{1.75}$$

$$\dot{\theta} = q \tag{1.76}$$

The dynamics of the vertical plane (denoted by subscript "v") as seen in (1.71)–(1.73) can be written in compact form as:

$$\mathbf{M}_\mathrm{v}\dot{\boldsymbol{\nu}}_\mathrm{v} + \mathbf{C}_\mathrm{v}(\boldsymbol{\nu}_\mathrm{v})\boldsymbol{\nu}_\mathrm{v} + \mathbf{D}_\mathrm{v}\boldsymbol{\nu}_\mathrm{v} + \mathbf{G}_\mathrm{v}(\boldsymbol{\eta}_\mathrm{v}) = \boldsymbol{\tau}_\mathrm{v} \tag{1.77}$$

where $\boldsymbol{\eta}_\mathrm{v} = \begin{bmatrix} x \; z \; \theta \end{bmatrix}^\mathrm{T}$, $\boldsymbol{\nu}_\mathrm{v} = \begin{bmatrix} u \; w \; q \end{bmatrix}^\mathrm{T}$, $\boldsymbol{\tau}_\mathrm{v} = \mathbf{T}_\mathrm{v}\mathbf{F}_\mathbf{T}\bar{\mathbf{u}} = \begin{bmatrix} \tau_x \; \tau_z \; \tau_\theta \end{bmatrix}^\mathrm{T}$, and

$$\mathbf{M}_\mathrm{v} = \begin{bmatrix} m - X_{\dot{u}} & 0 & 0 \\ 0 & m - Z_{\dot{w}} & 0 \\ 0 & 0 & I_y - M_{\dot{q}} \end{bmatrix}, \mathbf{C}_\mathrm{v}(\boldsymbol{\nu}_\mathrm{v}) = \begin{bmatrix} 0 & (m - Z_{\dot{w}})q & 0 \\ 0 & 0 & (X_{\dot{u}} - m)u \\ 0 & (Z_{\dot{w}} - X_{\dot{u}})u & 0 \end{bmatrix},$$

$$\mathbf{D}_\mathrm{v} = \begin{bmatrix} X_u & 0 & 0 \\ 0 & Z_w & 0 \\ 0 & 0 & M_q \end{bmatrix} \mathbf{T}_\mathrm{v} = \begin{bmatrix} 1 & 1 & 0 & 0 \\ 0 & 0 & \cos\beta & \cos\beta \\ -\varepsilon & -\varepsilon & \alpha\cos\beta & -\alpha\cos\beta \end{bmatrix} \mathbf{G}_\mathrm{v} = \begin{bmatrix} 0 \\ 0 \\ (z_G - z_B)W\sin\theta \end{bmatrix}$$

The kinematics equations of motion in the vertical plane (1.74)–(1.76) can be written as:

$$\dot{\boldsymbol{\eta}}_\mathrm{v} = \mathbf{J}_\mathrm{v}(\boldsymbol{\eta}_\mathrm{v})\boldsymbol{\nu}_\mathrm{v} \tag{1.78}$$

where

$$\mathbf{J}_\mathrm{v}(\boldsymbol{\eta}_\mathrm{v}) = \begin{bmatrix} \cos\theta & \sin\theta & 0 \\ -\sin\theta & \cos\theta & 0 \\ 0 & 0 & 1 \end{bmatrix}$$

Equations (1.77) and (1.78) can be rewritten in state-space form as:

$$\dot{\mathbf{x}}_{\mathbf{v}} = \bar{\mathbf{f}}_{\mathbf{v}}(\mathbf{x}_{\mathbf{v}}, t) + \bar{\mathbf{g}}_{\mathbf{v}}(\bar{\mathbf{u}}, t) \tag{1.79}$$

where

$$\bar{\mathbf{f}}_{\mathbf{v}}(\mathbf{x}_{\mathbf{v}}, t) = \begin{bmatrix} \mathbf{J}_{\mathbf{v}}(\boldsymbol{\eta}_{\mathbf{v}})\boldsymbol{\nu}_{\mathbf{v}} \\ -\mathbf{M}_{\mathbf{v}}^{-1}\left[(\mathbf{C}_{\mathbf{v}}(\boldsymbol{\nu}_{\mathbf{v}}) + \mathbf{D}_{\mathbf{v}})\boldsymbol{\nu}_{\mathbf{v}} + \mathbf{G}_{\mathbf{v}}(\boldsymbol{\eta}_{\mathbf{v}})\right] \end{bmatrix};$$

$$\bar{\mathbf{g}}_{\mathbf{v}}(\bar{\mathbf{u}}, t) = \begin{bmatrix} \mathbf{0}_{3\times 1} \\ \left[\, \bar{\mathbf{g}}_{\mathbf{v1}}\ \bar{\mathbf{g}}_{\mathbf{v2}}\ \bar{\mathbf{g}}_{\mathbf{v3}} \,\right]\mathbf{T}_{\mathbf{v}}\mathbf{F}_{\mathbf{T}}\bar{\mathbf{u}} \end{bmatrix}$$

$$\bar{\mathbf{g}}_{\mathbf{v1}} = \begin{bmatrix} \frac{1}{m-X_{\dot{u}}} & 0 & 0 \end{bmatrix}^{\mathrm{T}};\ \bar{\mathbf{g}}_{\mathbf{v2}} = \begin{bmatrix} 0 & \frac{1}{m-Z_{\dot{w}}} & 0 \end{bmatrix}^{\mathrm{T}};\ \bar{\mathbf{g}}_{\mathbf{v3}} = \begin{bmatrix} 0 & 0 & \frac{1}{I_y-M_{\dot{q}}} \end{bmatrix}^{\mathrm{T}} \tag{1.80}$$

1.7 Linear ROV Subsystem Model

It is useful to comprehend the open-loop dynamic properties of the ROV before closed-loop control system design. Since vast established tools have been present for linear system analysis, for preliminary analysis, a linearized model of the ROV at different operating conditions such as station-keeping conditions, horizontal and vertical planes are provided.

The linear equations of motion are obtained by linearization of the nonlinear ROV equations about a time-varying reference trajectory or an equilibrium point, for instance:

$$\boldsymbol{\nu}_{\mathbf{o}}(t) = [\, u_o(t)\ v_o(t)\ w_o(t)\ p_o(t)\ q_o(t)\ r_o(t) \,]^T \tag{1.80}$$

$$\boldsymbol{\eta}_{\mathbf{o}}(t) = [\, x_o(t)\ y_o(t)\ z_o(t)\ \phi_o(t)\ \theta_o(t)\ \psi_o(t) \,]^T \tag{1.81}$$

Let the perturbation from the reference trajectory $\boldsymbol{\nu}_{\mathbf{o}}(t)$ and $\boldsymbol{\eta}_{\mathbf{o}}(t)$ be described by the differentials:

$$\Delta\boldsymbol{\nu}(t) = \boldsymbol{\nu}(t) - \boldsymbol{\nu}_{\mathbf{o}}(t) \tag{1.82}$$

$$\Delta\boldsymbol{\eta}(t) = \boldsymbol{\eta}(t) - \boldsymbol{\eta}_{\mathbf{o}}(t) \tag{1.83}$$

$$\Delta\boldsymbol{\tau}_{\mathbf{A}}(t) = \boldsymbol{\tau}_{\mathbf{A}}(t) - \boldsymbol{\tau}_{\mathbf{Ao}}(t) \tag{1.84}$$

Introducing the following vector notation:

$$\mathbf{F_c}(t) = \mathbf{C}(\boldsymbol{\nu})\boldsymbol{\nu} \tag{1.85}$$

$$\mathbf{F_d}(t) = \mathbf{D}\boldsymbol{\nu} \tag{1.86}$$

implies that the ROV nonlinear equation can be linearized according to:

$$\mathbf{M}\Delta\dot{\boldsymbol{\nu}} + \left.\frac{d\mathbf{F_c}}{d\boldsymbol{\nu}}\right|_{\boldsymbol{\nu}_o} \Delta\boldsymbol{\nu} + \left.\frac{d\mathbf{F_d}}{d\boldsymbol{\nu}}\right|_{\boldsymbol{\nu}_o} \Delta\boldsymbol{\nu} + \left.\frac{d\mathbf{G}(\boldsymbol{\eta})}{d\boldsymbol{\eta}}\right|_{\boldsymbol{\eta}_o} \Delta\boldsymbol{\eta} = \Delta\boldsymbol{\tau}_\mathbf{A} \tag{1.87}$$

Perturbating the kinematic equations yields:

$$\dot{\boldsymbol{\eta}}_o + \Delta\dot{\boldsymbol{\eta}} = \mathbf{J}(\boldsymbol{\eta}_o + \Delta\boldsymbol{\eta})[\boldsymbol{\nu}_o + \Delta\boldsymbol{\nu}] \tag{1.88}$$

Substituting $\dot{\boldsymbol{\eta}}_o = \mathbf{J}(\boldsymbol{\eta}_o)\boldsymbol{\nu}_o$ into this expression implies that:

$$\Delta\dot{\boldsymbol{\eta}} = \mathbf{J}(\boldsymbol{\eta}_o + \Delta\boldsymbol{\eta})\Delta\boldsymbol{\nu} + \underbrace{[\mathbf{J}(\boldsymbol{\eta}_o + \Delta\boldsymbol{\eta}) - \mathbf{J}(\boldsymbol{\eta}_o)]\boldsymbol{\nu}_o}_{\mathbf{J}^*} \tag{1.89}$$

Linear theory implies that second-order terms can be neglected. Hence,

$$\Delta\dot{\boldsymbol{\eta}} = \mathbf{J}(\boldsymbol{\eta}_o)\Delta\boldsymbol{\nu} + \mathbf{J}^*(\boldsymbol{\nu}_o, \boldsymbol{\eta}_o)\Delta\boldsymbol{\eta} \tag{1.90}$$

Defining $\mathbf{x}_1 = \Delta\boldsymbol{\nu}$ and $\mathbf{x}_2 = \Delta\boldsymbol{\eta}$ yields the following linear time-varying model:

$$\mathbf{M}\dot{\mathbf{x}}_1 + \mathbf{C}(t)\mathbf{x}_1 + \mathbf{D}(t)\mathbf{x}_1 + \mathbf{G}(t)\mathbf{x}_2 = \boldsymbol{\tau}_\mathbf{A} \tag{1.91}$$

$$\dot{\mathbf{x}}_2 = \mathbf{J}(t)\mathbf{x}_1 + \mathbf{J}^*(t)\mathbf{x}_2 \tag{1.92}$$

where

$$\mathbf{C}(t) = \left.\frac{d\mathbf{F_c}}{d\boldsymbol{\nu}}\right|_{\boldsymbol{\nu}_o}$$

$$\mathbf{D}(t) = \left.\frac{d\mathbf{F_d}}{d\boldsymbol{\nu}}\right|_{\boldsymbol{\nu}_o}$$

$$\mathbf{G}(t) = \left.\frac{d\mathbf{G}(\boldsymbol{\eta})}{d\boldsymbol{\eta}}\right|_{\boldsymbol{\eta}_o}$$

$$\mathbf{J}(t) = \mathbf{J}(\boldsymbol{\eta}_o(t))$$

$$\mathbf{J}^*(t) = \mathbf{J}^*(\boldsymbol{\nu}_o(t), \boldsymbol{\eta}_o(t)) \tag{1.93}$$

Now, the linear state-space model of the ROV can be formulated. Defining $\mathbf{x} = \left[\mathbf{x}_1^{\mathrm{T}}\ \mathbf{x}_2^{\mathrm{T}}\right]^{\mathrm{T}}$, we obtain the following state-space model:

$$\begin{bmatrix} \dot{\mathbf{x}}_1 \\ \dot{\mathbf{x}}_2 \end{bmatrix} = \overbrace{\begin{bmatrix} -\mathbf{M}^{-1}[\mathbf{C}(t) + \mathbf{D}(t)] & -\mathbf{M}^{-1}\mathbf{G}(t) \\ \mathbf{J}(t) & \mathbf{J}^{*}(t) \end{bmatrix}}^{\mathbf{A}(t)} \begin{bmatrix} \mathbf{x}_1 \\ \mathbf{x}_2 \end{bmatrix} + \overbrace{\begin{bmatrix} \mathbf{M}^{-1} \\ \mathbf{0} \end{bmatrix}}^{\mathbf{B}} \boldsymbol{\tau}_{\mathbf{A}} \tag{1.94}$$

This can be written in abbreviated form as:

$$\dot{\mathbf{x}} = \mathbf{A}(t)\mathbf{x} + \mathbf{B}\tau_{\mathbf{A}} \tag{1.95}$$

In many ROV applications, it is reasonable to assume that the neutrally buoyant ($W = B$) ROV is moving in the vertical plane. In this condition, the steady-state linear and angular components are assumed as: $v_o = p_o = q_o = r_o = 0$ and that the equilibrium point is defined by the zero roll and pitch angles, that is $\phi_o = \theta_o = 0$. Hence, the time-varying matrices for the vertical plane motion (defined by certain u_o, w_o) with a fixed heading angle ψ_o is simplified to the following constant matrices:

$$\mathbf{M} = \begin{bmatrix} m - X_{\dot{u}} & 0 & 0 & 0 & 0 & 0 \\ 0 & m - Y_{\dot{v}} & 0 & 0 & 0 & 0 \\ 0 & 0 & m - Z_{\dot{w}} & 0 & 0 & 0 \\ 0 & 0 & 0 & I_{xx} - K_{\dot{p}} & -I_{xy} & -I_{xz} \\ 0 & 0 & 0 & -I_{xy} & I_{yy} - M_{\dot{q}} & -I_{yz} \\ 0 & 0 & 0 & -I_{xz} & -I_{yz} & I_{zz} - N_{\dot{r}} \end{bmatrix} \tag{1.96}$$

$$\mathbf{D} = -\begin{bmatrix} X_u & 0 & 0 & 0 & 0 & 0 \\ 0 & Y_v & 0 & 0 & 0 & 0 \\ 0 & 0 & Z_w & 0 & 0 & 0 \\ 0 & 0 & 0 & K_p & 0 & 0 \\ 0 & 0 & 0 & 0 & M_q & 0 \\ 0 & 0 & 0 & 0 & 0 & N_r \end{bmatrix} \tag{1.97}$$

$$\mathbf{C} = \begin{bmatrix} 0 & 0 & 0 & 0 & (m - Z_{\dot{w}})w_o & 0 \\ 0 & 0 & 0 & -(m - Z_{\dot{w}})w_o & 0 & (m - X_{\dot{u}})u_o \\ 0 & 0 & 0 & 0 & -(m - X_{\dot{u}})u_o & 0 \\ 0 & (m - Z_{\dot{w}})w_o & 0 & 0 & 0 & 0 \\ -(m - Z_{\dot{w}})w_o & 0 & (m - X_{\dot{u}})u_o & 0 & 0 & 0 \\ 0 & -(m - X_{\dot{u}})u_o & 0 & 0 & 0 & 0 \end{bmatrix} \tag{1.98}$$

$$\mathbf{G} = \begin{bmatrix} 0\,0\,0 & 0 & 0 & 0 \\ 0\,0\,0 & 0 & 0 & 0 \\ 0\,0\,0 & 0 & 0 & 0 \\ 0\,0\,0 & -z_B B & 0 & 0 \\ 0\,0\,0 & 0 & -z_B B & 0 \\ 0\,0\,0 & 0 & 0 & 0 \end{bmatrix} \tag{1.99}$$

If we assume that ψ_o is constant and $\phi_o = \theta_o = 0$, the kinematic transformation matrix **J** takes the form:

$$\mathbf{J} = \begin{bmatrix} \cos(\psi_o) & -\sin(\psi_o) & 0\,0\,0\,0 \\ \sin(\psi_o) & \cos(\psi_o) & 0\,0\,0\,0 \\ 0 & 0 & 1\,0\,0\,0 \\ 0 & 0 & 0\,1\,0\,0 \\ 0 & 0 & 0\,0\,1\,0 \\ 0 & 0 & 0\,0\,0\,1 \end{bmatrix} \tag{1.100}$$

whereas $\mathbf{J}^{*} = \mathbf{0}$. Consequently, the linear time-invariant model for both velocity $\mathbf{x_1}$ and position $\mathbf{x_2}$ can be written as:

$$\dot{\mathbf{x}} = \mathbf{A}\mathbf{x} + \mathbf{B}\tau_{\mathbf{A}} \tag{1.101}$$

or

$$\begin{bmatrix} \dot{\mathbf{x}}_1 \\ \dot{\mathbf{x}}_2 \end{bmatrix} = \begin{bmatrix} -\mathbf{M}^{-1}[\mathbf{C} + \mathbf{D}] & -\mathbf{M}^{-1}\mathbf{G} \\ \mathbf{J} & \mathbf{0} \end{bmatrix} \begin{bmatrix} \mathbf{x}_1 \\ \mathbf{x}_2 \end{bmatrix} + \begin{bmatrix} \mathbf{M}^{-1} \\ \mathbf{0} \end{bmatrix} \tau_{\mathbf{A}} \tag{1.102}$$

where **A** and **B** are constant matrices.

On the other hand, the time-varying matrices for the horizontal plane motion can be obtained by using the following constant **C** matrix (and the remaining matrices are identical to the case of the vertical plane motion):

$$\mathbf{C} = \begin{bmatrix} 0 & 0 & 0 & 0 & 0 & -(m - Y_{\dot{v}})v_o \\ 0 & 0 & 0 & 0 & 0 & (m - X_{\dot{u}})u_o \\ 0 & 0 & 0 & (m - Y_{\dot{v}})v_o & -(m - X_{\dot{u}})u_o & 0 \\ 0 & 0 & -(m - Y_{\dot{v}})v_o & 0 & (I_z - N_{\dot{r}})r_o & I_{yz}r_o \\ 0 & 0 & (m - X_{\dot{u}})u_o & -(I_z - N_{\dot{r}})r_o & 0 & -I_{xz}r_o \\ (m - Y_{\dot{v}})v_o & -(m - X_{\dot{u}})u_o & 0 & -I_{yz}r_o & I_{xz}r_o & 0 \end{bmatrix} \tag{1.103}$$

1.8 Summary

The dynamic and kinematics equations of the ROV were formulated. The mass inertia and buoyancy and gravitational force vector used in the dynamic equations, the hydrodynamic damping and added mass components can be obtained in Chap. 2. To provide for the study of the ROV dynamic behavior for subsequent control design, a simplified model for different planes and linear model of ROV was defined.

References

1. T.I. Fossen, *Marine Control Systems Guidance, Navigation, and Control of Ships* (Rigs and Underwater Vehicles, Marine Cybernetics, Trondheim, Norway, 2002)
2. C.S. Chin, Dynamic positioning simulation, thrust optimization design and control of a drillship under disturbances and faulty thruster. Simulation **88**(11), 1338–1349 (2012)

Chapter 2
Simulation of Remotely Operated Vehicle

Abstract The simulation consists of first selecting the software for simulation, examining the dynamic models, separating the model into several subsystems, testing the integrity of the entire model after combining the subsystems, and analyzing the results for it to make sense. In this chapter, we use the MATLAB™/Simulink™ software for simulation.

2.1 Mass Inertial with Added Mass Matrix

After having the ROV model, the next step is to perform simulation. The block diagrams are first separated into the following:

- Mass inertial with added mass matrix;
- Coriolis and centripetal with added mass matrix
- Hydrodynamic damping force matrix
- Buoyancy and gravitational force matrix
- Euler transformation
- Thruster configuration matrix.

The entire block diagram for ROV can be seen in Fig. 2.1, where each subsystem or block diagram can be constructed in a step-by-step manner to achieve the dynamics and kinematics of ROV. The important step in creating the block diagrams is to construct the "normalized" equation as seen in (2.2).

From (1.15), the dynamics of ROV can be written as:

$$\mathbf{M}\dot{\boldsymbol{\nu}} + \mathbf{C}(\boldsymbol{\nu})\boldsymbol{\nu} + \mathbf{D}(\boldsymbol{\nu})\boldsymbol{\nu} + \mathbf{G_f}(\boldsymbol{\eta}) = \boldsymbol{\tau}_\mathbf{A} \tag{2.1}$$

Manipulate (2.1) gives the normalized equation:

Electronic supplementary material The online version of this chapter (https://doi.org/10.1007/978-981-15-6511-3_2) contains supplementary material, which is available to authorized users.

C. S. Chin and M. W. S. Lau, *Benchmark Models of Control System Design for Remotely Operated Vehicles*, https://doi.org/10.1007/978-981-15-6511-3_2

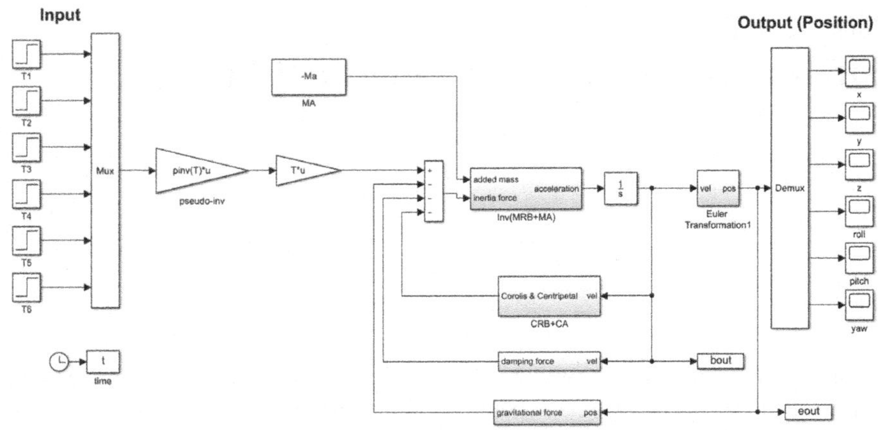

Fig. 2.1 Overall block diagram

$$\dot{\boldsymbol{\nu}} = \mathbf{M}^{-1}\{\boldsymbol{\tau}_\mathbf{A} - [\mathbf{C}(\boldsymbol{\nu})\boldsymbol{\nu} + \mathbf{D}(\boldsymbol{\nu})\boldsymbol{\nu} + \mathbf{G_f}(\boldsymbol{\eta})]\} \tag{2.2}$$

Now, we can construct the mass inertial with added mass matrix and its inverse as shown below.

The sum of the rigid body inertia mass and added fluid inertia mass matrix can be written as $\mathbf{M} = \mathbf{M_{RB}} + \mathbf{M_A} \in \Re^{6\times 6}$ where the individual terms in the inertia matrix in (1.7) are rewritten here.

$$\mathbf{M_{RB}} = \begin{bmatrix} m & 0 & 0 & 0 & mz_G & -my_G \\ 0 & m & 0 & -mz_G & 0 & mx_G \\ 0 & 0 & m & my_G & -mx_G & 0 \\ 0 & -mz_G & my_G & I_x & -I_{xy} & -I_{xz} \\ mz_G & 0 & -mx_G & -I_{yx} & I_y & -I_{yz} \\ -my_G & mx_G & 0 & -I_{zx} & -I_{zy} & I_z \end{bmatrix} \tag{2.3}$$

and the hydrodynamic added mass matrix can be seen as below:

$$\mathbf{M_A} = \begin{bmatrix} X_{\dot{u}} & X_{\dot{v}} & X_{\dot{w}} & X_{\dot{p}} & X_{\dot{q}} & X_{\dot{r}} \\ Y_{\dot{u}} & Y_{\dot{v}} & Y_{\dot{w}} & Y_{\dot{p}} & Y_{\dot{q}} & Y_{\dot{r}} \\ Z_{\dot{u}} & Z_{\dot{v}} & Z_{\dot{w}} & Z_{\dot{p}} & Z_{\dot{q}} & Z_{\dot{r}} \\ K_{\dot{u}} & K_{\dot{v}} & K_{\dot{w}} & K_{\dot{p}} & K_{\dot{q}} & K_{\dot{r}} \\ M_{\dot{u}} & M_{\dot{v}} & M_{\dot{w}} & M_{\dot{p}} & M_{\dot{q}} & M_{\dot{r}} \\ N_{\dot{u}} & N_{\dot{v}} & N_{\dot{w}} & N_{\dot{p}} & N_{\dot{q}} & N_{\dot{r}} \end{bmatrix} \tag{2.4}$$

The rigid body inertia mass $\mathbf{M_{RB}} \in \Re^{6\times 6}$ can be included as a matrix input. The matrix input can be applied to the added fluid inertia mass matrix, $\mathbf{M_A} \in \Re^{6\times 6}$.

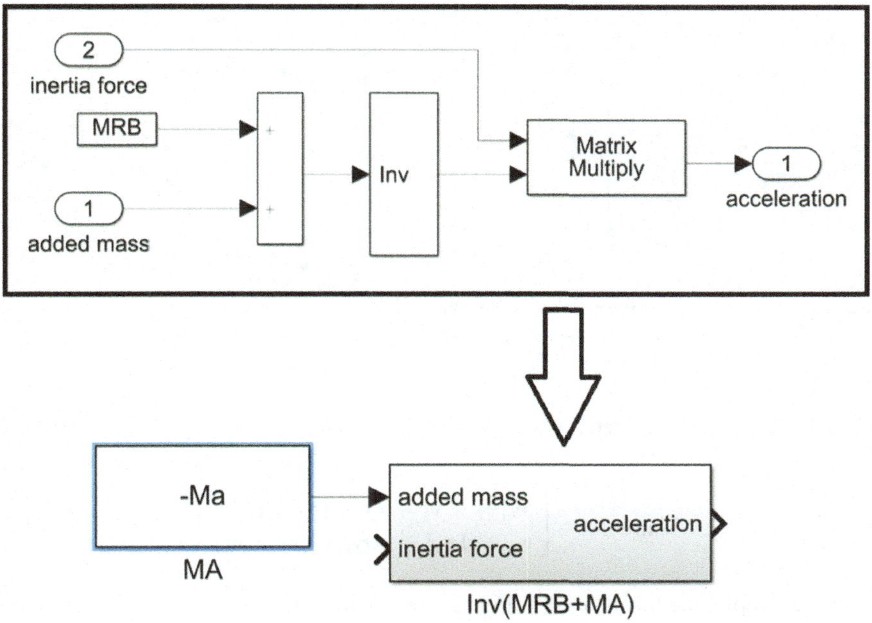

Fig. 2.2 Mass inertial and added mass block diagram

The inverse block is used to perform the inverse of the ($\mathbf{M} = \mathbf{M_{RB}} + \mathbf{M_A}$) before multiplying by the inertia forces to obtain the acceleration, as seen in Fig. 2.2.

2.2 Coriolis and Centripetal with Added Mass Matrix

The following diagram is an illustration of the Coriolis and centripetal with added mass matrix, $\mathbf{C}(\boldsymbol{\nu}) = \mathbf{C_{RB}}(\boldsymbol{\nu}) + \mathbf{C_A}(\boldsymbol{\nu}) \in \Re^{6\times6}$ as seen below. It consists of the sum of Coriolis and centripetal and its added mass matrix as seen in Fig. 2.3.

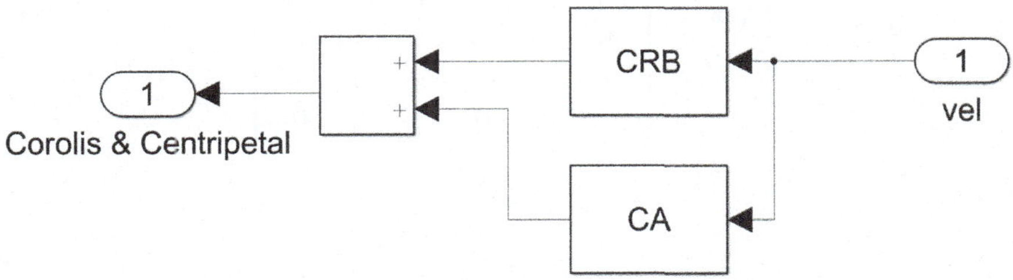

Fig. 2.3 Coriolis and centripetal matrix with added mass block diagram

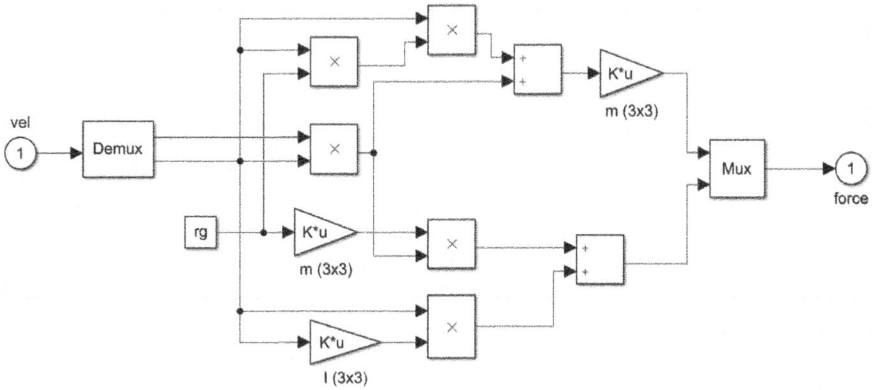

Fig. 2.4 Coriolis and centripetal matrix (C_{RB}) block diagram

$$\mathbf{C_{RB}}(\mathbf{v})\mathbf{v} = \begin{bmatrix} m[\mathbf{v_2} \times \mathbf{v_1} + \mathbf{v_2} \times (\mathbf{v_2} \times \mathbf{r_G})] \\ \mathbf{v_2} \times (\mathbf{I}\mathbf{v_2}) + m\mathbf{r_G} \times (\mathbf{v_2} \times \mathbf{v_1}) \end{bmatrix} \tag{2.5}$$

For example, the terms used in Fig. 2.4 are as follows.

- Circle(In) is the input of the subsystem
- Circle(Out) is the output of the subsystem
- Block "rg" carries the value of the location of the center of gravity of ROV
- Block "X" simply represent multiplication of two inputs
- Block "+ +" simply represents the addition of two inputs
- Triangle "K*u" means that the input is multiplied by a mass matrix of ROV denoted by "m"
- Triangle "I*u" means that the input is multiplied by inertia matrix of ROV denoted by "I".

The added mass terms for the Coriolis and centripetal can be rewritten as:

$$\mathbf{C_A}(\mathbf{v}) = \begin{bmatrix} 0 & 0 & 0 & 0 & -a_3 & a_2 \\ 0 & 0 & 0 & a_3 & 0 & -a_1 \\ 0 & 0 & 0 & -a_2 & a_1 & 0 \\ 0 & -a_3 & a_2 & 0 & -b_3 & b_2 \\ a_3 & 0 & -a_1 & b_3 & 0 & -b_1 \\ -a_2 & a_1 & 0 & -b_2 & b_1 & 0 \end{bmatrix} \tag{2.6}$$

where

$$\begin{aligned} a_1 &= X_{\dot{u}}u + X_{\dot{v}}v + X_{\dot{w}}w + X_{\dot{p}}p + X_{\dot{q}}q + X_{\dot{r}}r \\ a_2 &= X_{\dot{v}}u + Y_{\dot{v}}v + Y_{\dot{w}}w + Y_{\dot{p}}p + Y_{\dot{q}}q + Y_{\dot{r}}r \\ a_3 &= X_{\dot{w}}u + Y_{\dot{w}}v + Z_{\dot{w}}w + Z_{\dot{p}}p + Z_{\dot{q}}q + Z_{\dot{r}}r \end{aligned}$$

$$\begin{aligned} b_1 &= X_{\dot{p}}u + Y_{\dot{p}}v + Z_{\dot{p}}w + K_{\dot{p}}p + K_{\dot{q}}q + K_{\dot{r}}r \\ b_2 &= X_{\dot{q}}u + Y_{\dot{q}}v + Z_{\dot{q}}w + K_{\dot{q}}p + M_{\dot{q}}q + M_{\dot{r}}r \\ b_3 &= X_{\dot{r}}u + Y_{\dot{r}}v + Z_{\dot{r}}w + K_{\dot{r}}p + M_{\dot{r}}q + N_{\dot{r}}r \end{aligned} \tag{2.7}$$

For example, the block named “a1” as seen in Fig. 2.5 is made up of added mass components as seen in Fig. 2.6.

The block on the top right-hand side of Fig. 2.5 is made up of added mass components as seen in the first row of (2.6) (Fig. 2.7).

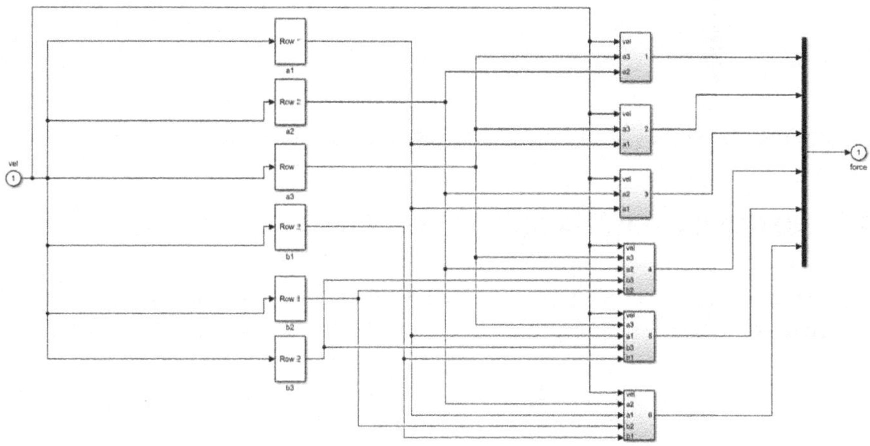

Fig. 2.5 Coriolis and centripetal added matrix (C_A) block diagram

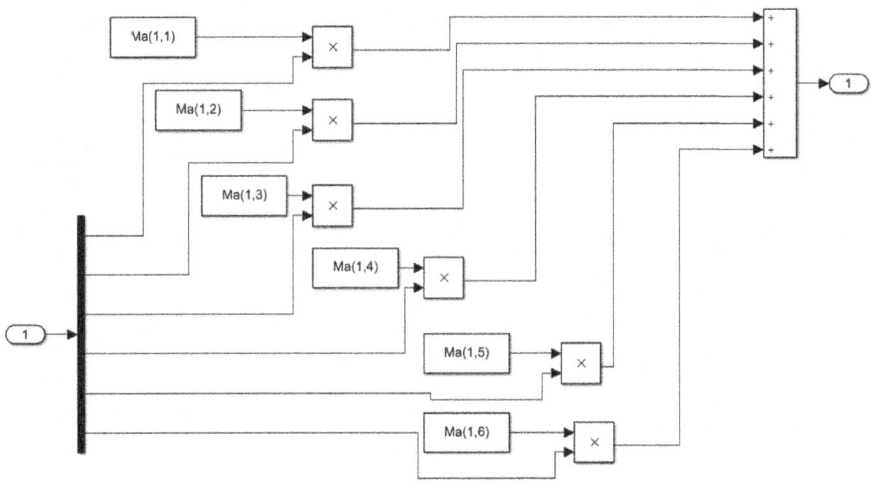

Fig. 2.6 Block named “a1” in Coriolis and centripetal added matrix (C_A)

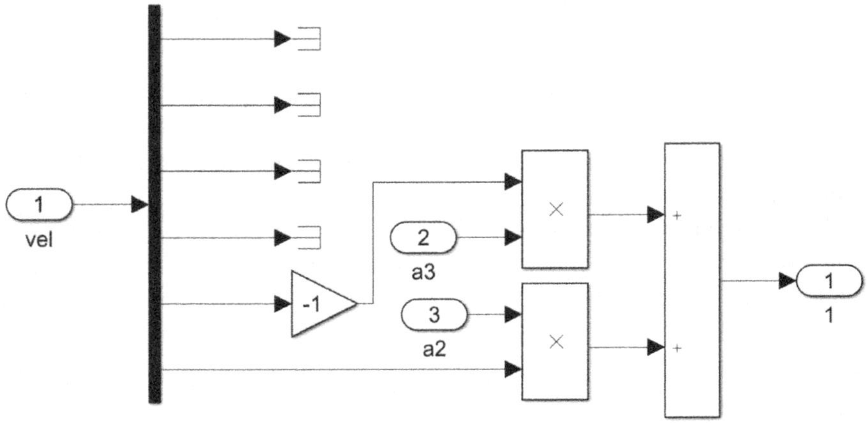

Fig. 2.7 First row of Coriolis and centripetal added matrix (C_A)

2.3 Hydrodynamic Damping Force Matrix

As seen in previous section, the hydrodynamic damping in the diagonal form can be modeled as follows:

$$\mathbf{D_L} = -\text{diag}\{X_u, Y_v, Z_w, K_p, M_q, N_r\} \tag{2.8a}$$

$$\mathbf{D_Q} = -\text{diag}\{X_{u|u|}, Y_{v|v|}, Z_{w|w|}, K_{p|p|}, M_{q|q|}, N_{r|r|}\} \tag{2.8b}$$

The block diagram for the hydrodynamic damping matrix can be seen in Fig. 2.8, where one can toggle between linear and quadratic hydrodynamic damping.

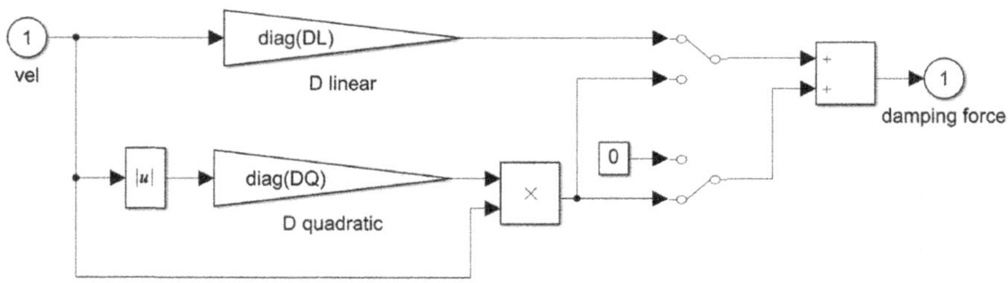

Fig. 2.8 Hydrodynamic damping force matrix block diagram

2.4 Buoyancy and Gravitational Force Vector

The buoyancy and gravitational force vector can be expressed as shown.

$$\mathbf{G_f}(\boldsymbol{\eta}) = \begin{bmatrix} (W-B)\sin\theta \\ -(W-B)\cos\theta\sin\phi \\ -(W-B)\cos\theta\cos\phi \\ -(y_G W - y_B B)\cos\theta\cos\phi + (z_G W - z_B B)\cos\theta\sin\phi \\ (z_G W - z_B B)\sin\theta + (x_G W - x_B B)\cos\theta\cos\phi \\ -(x_G W - x_B B)\cos\theta\sin\phi - (y_G W - y_B B)\sin\theta \end{bmatrix} \tag{2.9}$$

The block diagram can be formulated as seen in Fig. 2.9. The value from the input passes through the Demux into respective rows in (2.9) and combined into a single column output vector by the Mux.

As ROV can become neutrally buoyant, $W = B$ and placing an additional mass on the ROV to make the X–Y coordinates of the C.B. coincide with the X–Y coordinate of the CG ($x_G = x_B = 0$, $y_G = y_B = 0$), the resulting vector can become

$$\mathbf{G_f}(\boldsymbol{\eta}) = \begin{bmatrix} 0 \\ 0 \\ 0 \\ (z_G - z_B)W\cos\theta\sin\phi \\ (z_G - z_B)W\sin\theta \\ 0 \end{bmatrix} \tag{2.10}$$

The block diagram can be simplified into Fig. 2.10.

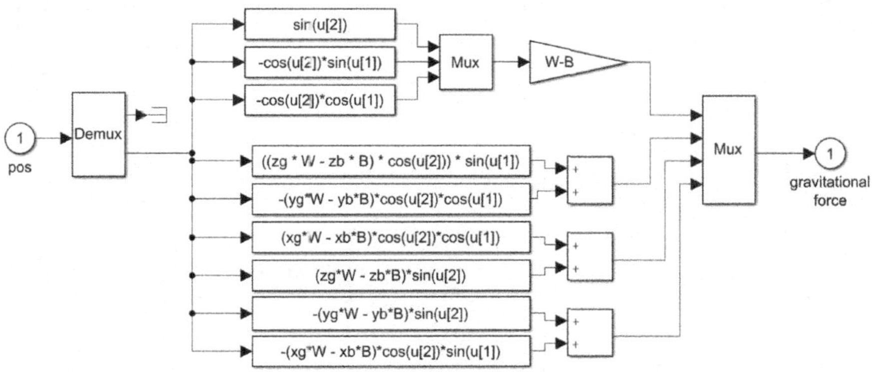

Fig. 2.9 Buoyancy and gravitational force vector block diagram

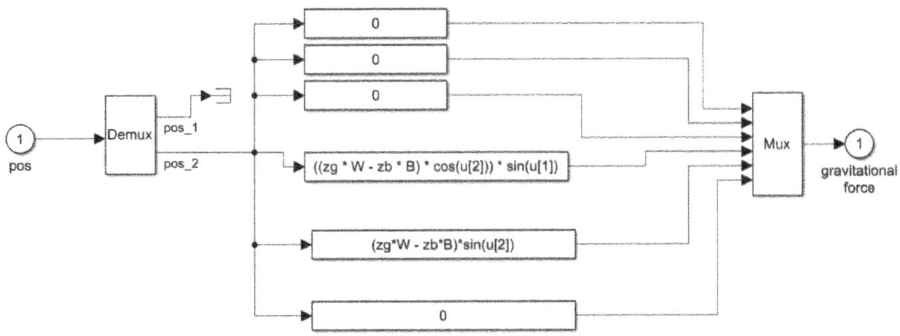

Fig. 2.10 Simplified buoyancy and gravitational force vector block diagram

2.5 Euler Transformation

The Euler transformation is used to transform the body-fixed coordinates to earth-fixed coordinates. There are six different velocity inputs for different directions, namely: surge (x), sway (y), heave (z), roll (ϕ), pitch(θ), and yaw(ψ). The velocity is therefore converted into two transformations, namely: Transformation T($\mathbf{J_1}$) and Transformation R ($\mathbf{J_2}$) as seen in Fig. 2.11.

$$\mathbf{J}(\eta) = \begin{bmatrix} \mathbf{J}_1(\eta_2) & 0 \\ 0 & \mathbf{J}_2(\eta_2) \end{bmatrix} \tag{2.11}$$

where

$$\mathbf{J}_1(\eta_2) = \begin{bmatrix} c(\psi)c(\theta) & -s(\psi)c(\phi) + c(\psi)s(\theta)s(\phi) & s(\psi)s(\phi) + c(\psi)c(\phi)s(\theta) \\ s(\psi)c(\theta) & c(\psi)c(\phi) + s(\phi)s(\theta)s(\psi) & -c(\psi)s(\phi) + s(\theta)s(\psi)c(\theta) \\ -s(\theta) & c(\theta)s(\phi) & c(\theta)c(\phi) \end{bmatrix} \tag{2.12}$$

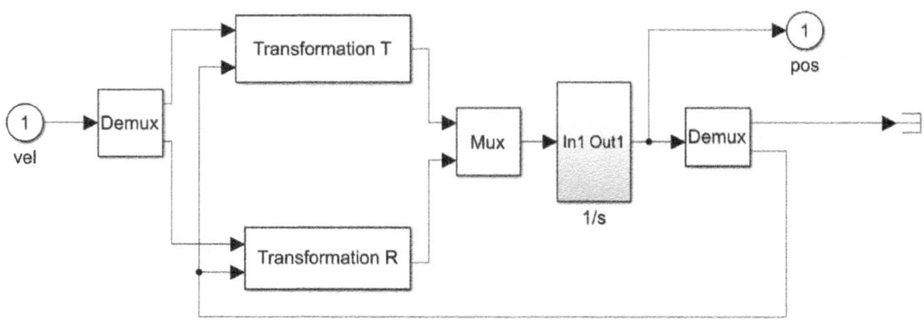

Fig. 2.11 Euler transformation block diagram

$$\mathbf{J_2}(\eta_2) = \begin{bmatrix} 1 & s(\phi)t(\theta) & c(\phi)t(\theta) \\ 0 & c(\phi) & -s(\phi) \\ 0 & \frac{s(\phi)}{c(\theta)} & \frac{c(\phi)}{c(\theta)} \end{bmatrix} \tag{2.13}$$

and s = sin(•), c = cos(•), t = tan(•).

For example, the first row of $\mathbf{J_1}$ is set using a Function Block Row 1, 2, and 3 as seen in Fig. 2.12. The parameters for the first row (see Fig. 2.13) are as follows: cos(u[3]) * cos(u[2]). The "[3]" refers to the input yaw or heading angle from the angular position. The "[2]" refers to the input pitch angle. The "[1]" refers to the input "roll" from the angular position. The same approach applies to Rows 2 and 3 as seen in Figs. 2.14 and 2.15, respectively.

For example, the first row of $\mathbf{J_2}$ is set using a Function Block Row 1, 2, and 3 as seen in Fig. 2.16. The parameter for the first row (see Fig. 2.17) is just 1. The same approach applies to Rows 2 and 3 as seen in Figs. 2.18 and 2.19, respectively.

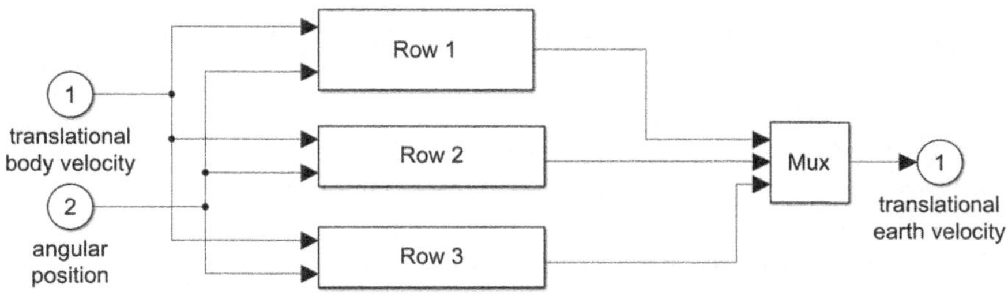

Fig. 2.12 Euler transformation T block diagram

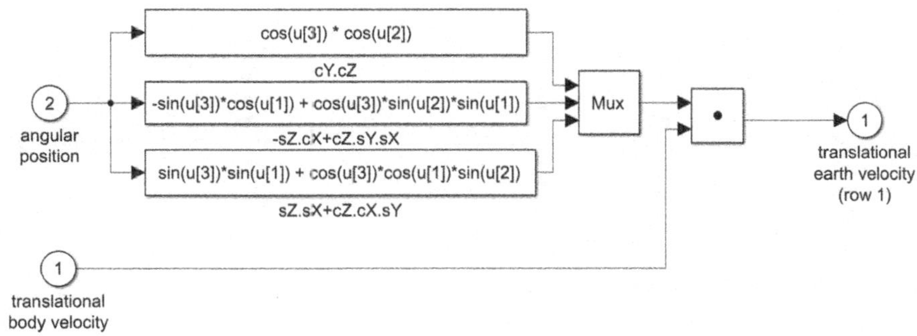

Fig. 2.13 Euler transformation T (Row 1) block diagram

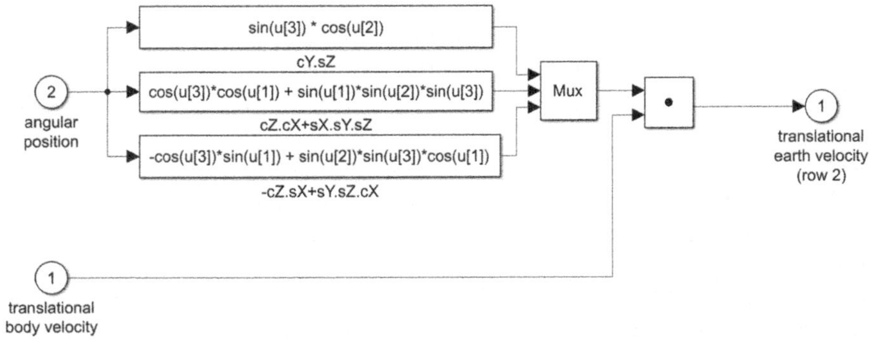

Fig. 2.14 Euler transformation T (Row 2) block diagram

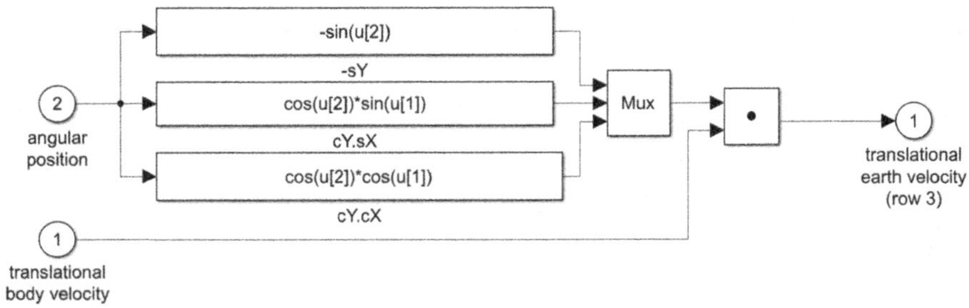

Fig. 2.15 Euler transformation T (Row 3) block diagram

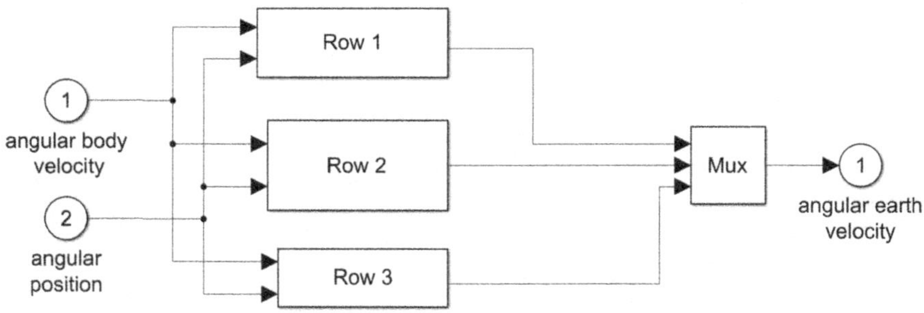

Fig. 2.16 Euler transformation R block diagram

The Euler transformation is the last process before the positions (output) are separated into their respecting rows through the Demux Block as shown in Fig. 2.20.

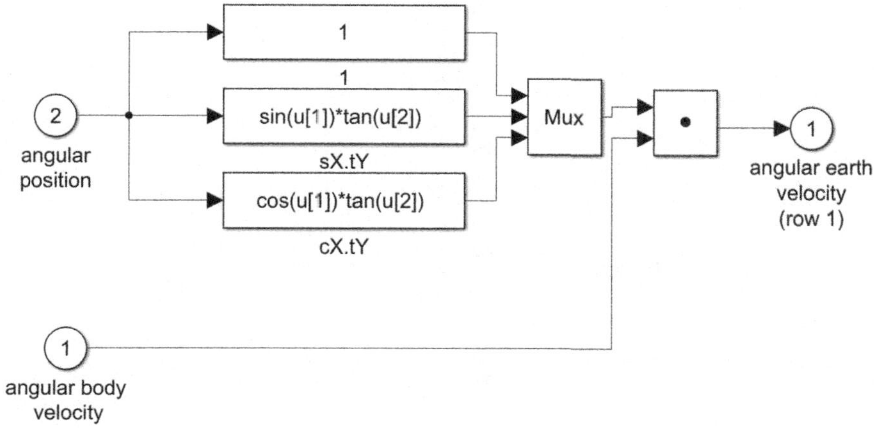

Fig. 2.17 Euler transformation R (Row 1) block diagram

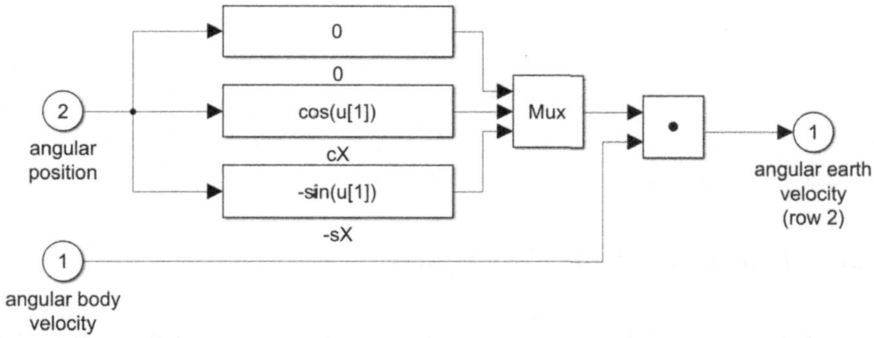

Fig. 2.18 Euler transformation R (Row 2) block diagram

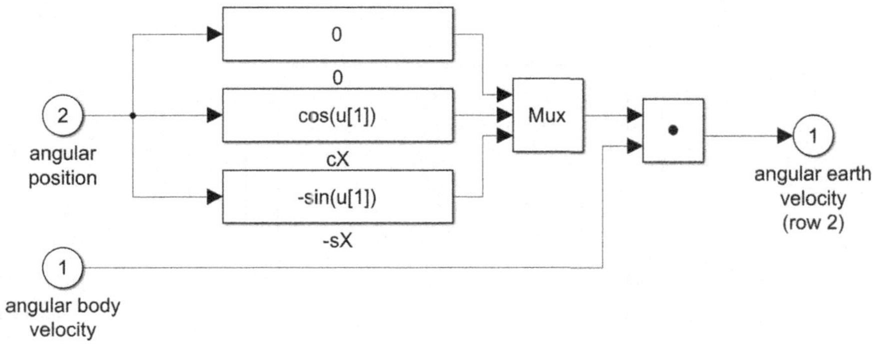

Fig. 2.19 Euler transformation R (Row 3) block diagram

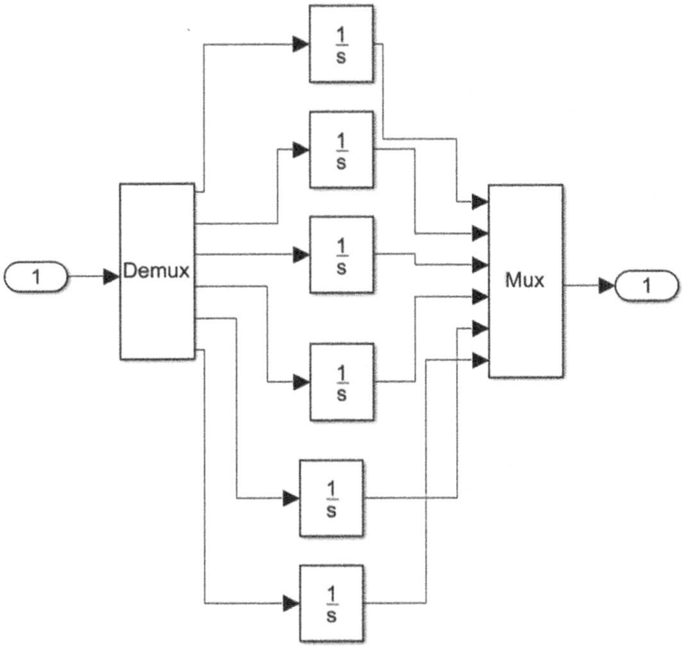

Fig. 2.20 Euler transformation integrator block diagram

2.6 Thruster Configuration Matrix

The thrusters' configuration matrix **T** can be modeled by a constant matrix as shown below. It represents the relationships between the input **u** and force and moment vector generated in the six degrees of freedom,

$$\begin{bmatrix} \tau_x \\ \tau_y \\ \tau_z \\ \tau_\phi \\ \tau_\theta \\ \tau_\psi \end{bmatrix} = \overbrace{\begin{bmatrix} 1 & 1 & 0 & 0 \\ 0 & 0 & \sin\beta & -\sin\beta \\ 0 & 0 & \cos\beta & \cos\beta \\ 0 & 0 & \delta & -\delta \\ -\varepsilon & -\varepsilon & \alpha\cos\beta & -\alpha\cos\beta \\ \gamma & -\gamma & \alpha\sin\beta & -\alpha\sin\beta \end{bmatrix}}^{\mathbf{T}} \overbrace{\begin{bmatrix} \mathrm{u}_1 \\ \mathrm{u}_2 \\ \mathrm{u}_3 \\ \mathrm{u}_4 \end{bmatrix}}^{\mathbf{u}}$$

$$\boldsymbol{\tau}_{\mathbf{A}} = \mathbf{T}\mathbf{u} \tag{2.14}$$

Since $\mathbf{u} = \mathbf{T}^* \boldsymbol{\tau}_{\mathbf{A}}$, the control input vector **u** can be computed as:

$$\mathbf{u} = \mathbf{T}^* \boldsymbol{\tau}_{\mathbf{A}} \tag{2.15}$$

where the Moore–Penrose pseudoinverse ("pinv") can be written as:

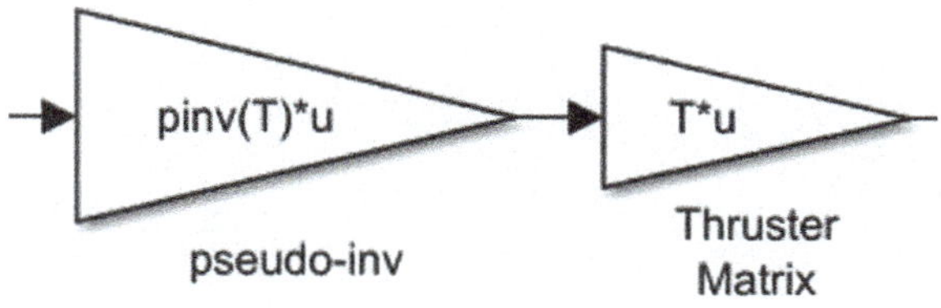

Fig. 2.21 Thruster matrix block diagram

$$\mathbf{T}^{*} = \mathbf{T}^{\mathrm{T}}(\mathbf{T}\mathbf{T}^{\mathrm{T}})^{-1} \tag{2.16}$$

The block diagram of the thruster configuration matrix can be seen in Fig. 2.21.

2.7 Simulation of ROV

The solver options used are fixed as follows: variable-step type, ODE45 (Dormand-Prince solver and relative tolerance of 0.001). The configuration parameter settings can be seen in Fig. 2.22. The entire block diagram of ROV can be constructed as shown in Fig. 2.23. The Simulink^{TM} model in Fig. 2.23 includes the output scopes to display the position responses. The output velocities and positions are saved as bout and eout, respectively. The thrusters T3 and T4 are set as 2 V, respectively.

The MATLAB^{TM} script to plot the velocity and position responses is shown in Figs. 2.24 and 2.25, respectively. As shown in Fig. 2.25, the ROV is unable to reach steady-state values.

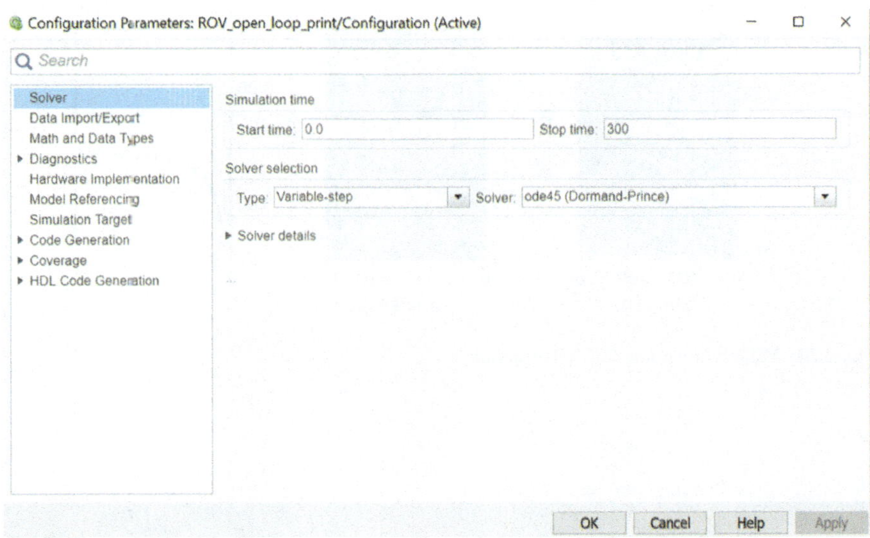

Fig. 2.22 Configuration parameters for ROV simulation

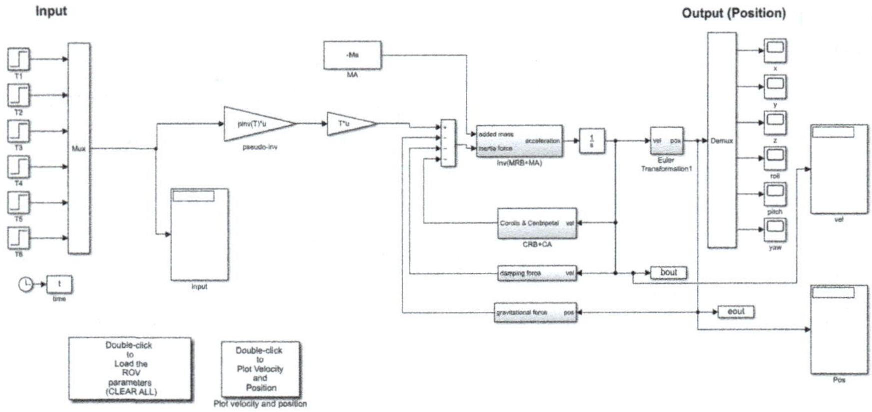

Fig. 2.23 ROV model for simulation

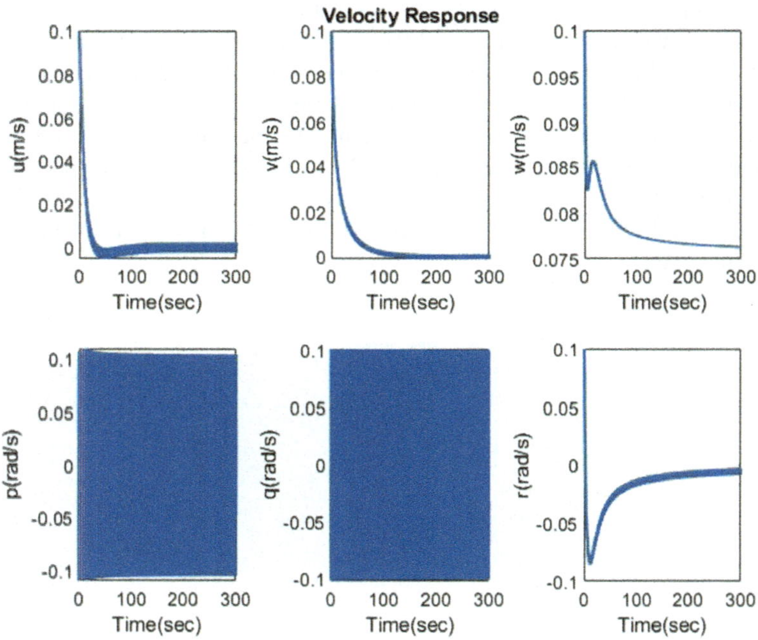

Fig. 2.24 Velocity responses of ROV model

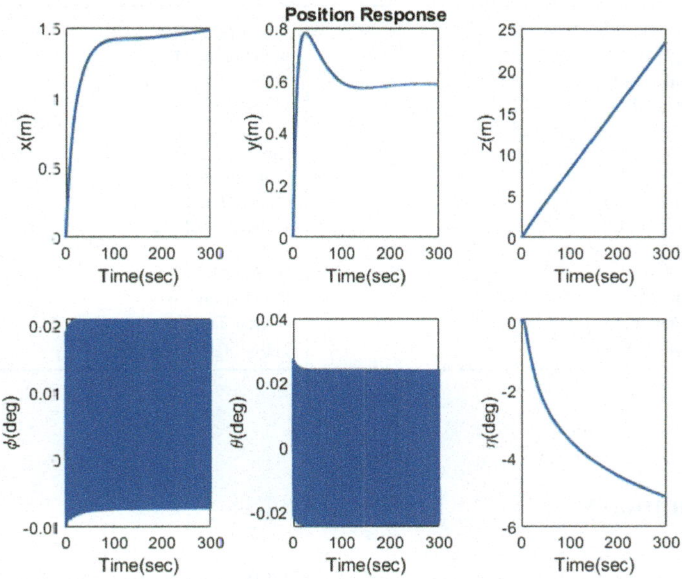

Fig. 2.25 Position responses of ROV model

% MATLAB Script figure(1) subplot(2,3,1) plot(t,eout(:,1)) xlabel('Time(sec)') ylabel('x(m)') subplot(2,3,2) plot(t,eout(:,2)) title('Position Response') xlabel('Time(sec)') ylabel('y(m)') subplot(2,3,3) plot(t,eout(:,3)) xlabel('Time(sec)') ylabel('z(m)')	figure(2) subplot(2,3,1) plot(t,bout(:,1)) xlabel('Time(sec)') ylabel('u(m/s)') subplot(2,3,2) plot(t,bout(:,2)) title('Velocity Response') xlabel('Time(sec)') ylabel('v(m/s)') subplot(2,3,3) plot(t,bout(:,3)) xlabel('Time(sec)') ylabel('w(m/s)')

subplot(2,3,4) plot(t,eout(:,4)) xlabel('Time(sec)') ylabel('\phi(rad)') subplot(2,3,5) plot(t,eout(:,5)) xlabel('Time(sec)') ylabel('\theta(rad)') subplot(2,3,6) plot(t,eout(:,6)) xlabel('Time(sec)') ylabel('\eta(rad)')	subplot(2,3,4) plot(t,bout(:,4)) xlabel('Time(sec)') ylabel('p(rad/s)') subplot(2,3,5) plot(t,bout(:,5)) xlabel('Time(sec)') ylabel('q(rad/s)') subplot(2,3,6) plot(t,bout(:,6)) xlabel('Time(sec)') ylabel('r(rad/s)')

2.8 Summary

The dynamic and kinematics equations of the ROV were simulated. The following block diagrams were modeled systematically:

- Mass inertial with added mass matrix;
- Coriolis and centripetal with added mass matrix;
- Hydrodynamic damping force matrix;
- Buoyancy and gravitational force matrix;
- Euler transformation;
- Thruster configuration matrix.

The details of the parameters used in the ROV model will be discussed in Chap. 3. The ROV model can be downloaded via the website:

- https://drive.google.com/open?id=1KkN-LHZ6H-TeIm-K9aaXWIIFtab2jvvA.

Chapter 3
Identification of Remotely Operated Vehicle and Benchmark ROV Model

Abstract The material discussed here on methods of determining the hydrodynamic parameters is only a small fraction of the enormous amount of works done and literature available. It is a short summary and includes the work done by the authors, literature published, and some of the papers referenced during the works. The primary goal is to estimate the model and the associated parameters for the analysis of the ROV performance and model-based control of the vehicle.

3.1 Laboratory-Based Method Using Tow Tank Facilities

The laboratory-based methods using physical model experiments can be expensive and elaborate in setup, the equipment, and measurement systems required, but they have provided a platform for the design and development of many classes of marine vessels for different types of environment. A fluid tank is a must in lab-based experimental methods, but the size of the tank and the equipment varies from simple, small, low cost to large with elaborate and controlled actuation, accurate and sophisticated instrumentations, data acquisitions, and data extractions.

Computational methods, requiring the case of computational fluid dynamics software, are now available and provide another way of determining these parameters. The approach uses computer-aided design tools to "build" the model of the vessel and allows users to perform computations and virtual experiments like lab tests.

Eventually, the model and the values obtained are either tested using computer model simulation or/and experimental trials in pools, tanks, or in the sea. Onboard sensors are used to collect data to validate the model and the coefficients.

The typical ROV model [1] as shown in Chaps. 1 and 2 are given as:

$$(\mathbf{M_{RB}} + \mathbf{M_A})\dot{\boldsymbol{\nu}} + [\mathbf{C_{RB}}(\boldsymbol{\nu}) + \mathbf{C_A}(\boldsymbol{\nu})]\boldsymbol{\nu} + \mathbf{D}(\boldsymbol{\nu})\boldsymbol{\nu} + \mathbf{G_f}(\boldsymbol{\eta}) = \boldsymbol{\tau}_{\mathbf{A}} + \boldsymbol{\tau}_{\mathbf{Environment}} \tag{3.1}$$

where the $\boldsymbol{\tau}_{\mathbf{Environment}}$ can be included as the environmental forces due to the wind, wave, and underwater current. The remaining parameter descriptions can be found in the previous chapters.

C. S. Chin and M. W. S. Lau, *Benchmark Models of Control System Design for Remotely Operated Vehicles*, https://doi.org/10.1007/978-981-15-6511-3_3

The main parameters to identify are the added mass $\mathbf{M_A}$, the Coriolis and centripetal terms due to rigid body $\mathbf{C_{RB}}(\mathbf{n})$, and due to added mass, $\mathbf{C_A}(\boldsymbol{\nu})$, the drag (resistance) $\mathbf{D}(\boldsymbol{\nu})$ term which comprises the linear and the quadratic drag forces, and buoyancy and gravitational force $\mathbf{G_f}(\boldsymbol{\eta})$. The rigid body mass $\mathbf{M_{RB}}$ can be obtained from CAD drawings.

Depending on the profile, shape, and the operational speeds, simplifications are made, reducing the effort and cost of identifying those parameters. For example, at low speed, $\mathbf{C}(\boldsymbol{\nu})\boldsymbol{\nu} \approx 0$, and $\mathbf{D}(\boldsymbol{\nu})\boldsymbol{\nu}$ comprises only the linear drag force.

Tow tank facilities are typically used to determine the forces and hydrodynamic parameters of marine vessels. The method is one of the oldest. The planar motion mechanism (PMM) is an important piece of equipment in the Tow tank lab and is used to generate steady-state and controlled motions, to perform experimental studies with ship models to determine the maneuvering characteristics of a marine vessel. Typically, a scaled model of the target marine vessel is mounted onto a PMM or a driven platform that carries a host sensor to measure reaction forces and the resulting motions. The platform can be driven in predefined trajectories in one, two, or all of the forced displacements and at various speeds, and various types of disturbances can be introduced during the motion. Measured applied forces and moments and post data analysis are used to extract the hydrodynamic characteristics of the vehicles and to provide numerical values for hydrodynamic stability derivatives needed to quantify the directional stability, for example, vessel's straight-line stability and vessel's controllability. To define all of the stability derivatives, both static tests and dynamic tests are required.

One issue with the use of this method is that it can be costly and time-consuming. However, it is still one of the methods capable of identifying coefficients in many degrees of freedom.

A low-cost tank test determination of added mass and drag coefficient was presented [2] using the chain and pulley system. This setup is quite like the historical model tests done in the Peerlesspool in London in 1761. It can produce constant speed. Displacement was produced when known external weight attached to a system of chain and pulley was released from rest and measured by an encoder on one of the pulleys. The model of interest in the x-direction is of the form

$$M\ddot{x} + 0.5\rho C_d A\dot{x}^2 = F_0 + \Delta m_w g \tag{3.2}$$

where F_0 includes the buoyancy and gravitational force, ρ is the density of water, ρ is the cross-sectional area of the object, Δm_w is mass of the additional external weight attached to the pulley and M includes the added mass term m_a, the pulley mass m_p and external weight mass m_w. This equation has the same form as that used to solve for the terminal velocity of a free-falling object of mass m in x-direction:

$$m\frac{d^2x}{dt^2} + k\left(\frac{dx}{dt}\right)^2 = mg \tag{3.3}$$

where closed-form solutions for displacement and speed can be found. The novelty of their method is the way they numerically solved the closed-form solutions to obtain the added mass and the drag coefficients.

3.2 Free Decay Method

Another tank test without the use of the PMM is the method based on free decay motion analysis, which has been used for the study of marine vessels as mentioned in the book [3] (see Sect. 7.1.4, 1990). The free decay test has often been used to determine mechanical parameters like natural frequency and damping and is done by displacing the system from its equilibrium position and allowing it to oscillate freely in the media. Sensors and vision systems are used to measure the displacements and using least-square method (as suggested by [4], for example) those parameters are determined independently.

A 1-D free decay method [5] was described. The method suspended the vehicle from an overhead crane with springs that were "sized" to produce the appropriate oscillatory amplitude so that the following equation correctly describes the forces acting on the vehicle in an oscillatory regime.

$$(m + m_a)\ddot{x} + 0.5\rho C_d A\dot{x}|\dot{x}| = F \tag{3.4}$$

Ross et al. [6] extended the method to determine hydrodynamic parameters in more degrees of freedom—the longitudinal and lateral degrees of freedom—by suspending the vehicle by linear springs (see Fig. 3.1), and measuring the generalized displacements and applying differentiation filtering techniques to generate the generalized derivatives, accelerations, and speed for use in the least-square fit algorithms. The parameters are used with the respective mathematical models, and the simulation results are compared favorably with the measured generalized coordinates.

Fig. 3.1 Free decay approach by using 1-D linear

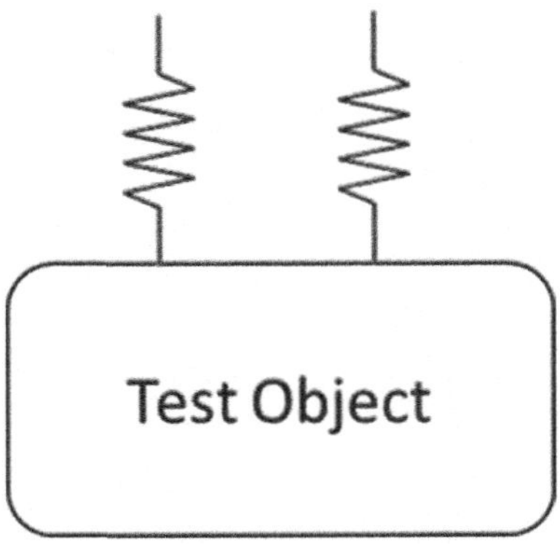

3.3 Free Decay Pendulum Method

A variation in the free decay method for low-speed small-size vehicle [7] was proposed. Instead of linear springs, a scaled model of the vehicle was attached to a rod pivoted on bearings—like a mass on the end of a pendulum as seen in Fig. 3.2. It could oscillate freely when displaced from the equilibrium position.

This method was used to measure the surge and sway motions, and by the use of torsional spring, the free decay oscillations in the yaw direction are measured. It is a 1-D modeling. For example, the model in the body frame motion in x-direction is given as

$$\ddot{x} = \frac{B - mg}{(m + m_a)} \sin\theta - \frac{D_L}{(m + m_a)}\dot{x} - \frac{D_Q}{(m + m_a)}\dot{x}|\dot{x}| \tag{3.5}$$

A video camera was used to record the motion, and the image is digitized to obtain displacements, and by numerical differentiations, speed and acceleration are obtained. Data can be corrupted by noise unless some filtering method is used.

The results for the added mass m_a are quite consistent with the values found from the computational software WAMIT. However, the results for damping forces were not so crisp. The damping term comprising the linear damping term and the quadratic damping term is difficult to separate out these two effects

$$D(|v|) = D_L + D_Q(|v|) \tag{3.6}$$

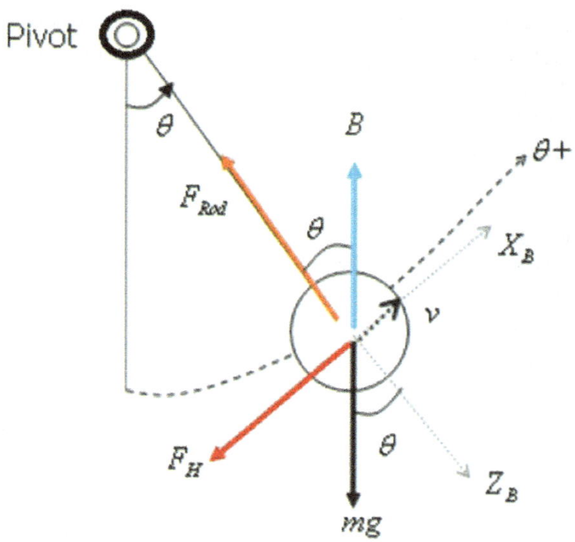

Fig. 3.2 Free decay pendulum method [7]

The term by term comparison of the estimate of the linear and quadratic drag force with ANSYS CFX was not so good, but the total of the drag forces compared very well with the corresponding sum of the drag forces from ANSYS CFX.

A detailed analysis and discussion on the free decay pendulum method was presented [8] using a spherical ball instead of a vehicle model. This allows for simple and direct comparisons with theories based on ideal and inviscid flow. Instead of using derivatives of the displacement data, they devised a method of numerical integration of the assumed model and the estimate of the parameters that minimizes the errors between the generated displacements and the measured displacements, thereby avoiding the effect of noisy data due to differentiation. A detailed analysis using the theory based on a sphere shows that the added mass data obtained by the free decay oscillation is more trustworthy than computational methods using Navier–Stokes equations based on the idealized, inviscid flow. Their study also showed that discriminating between the linear and the quadratic drag is difficult. A different method may be necessary to better identify the linear and quadratic coefficient.

An interesting note by Gartner et al. on the free decay pendulum method is the possible effect on the motion of the model when it reverses and moves into the wake of the forward direction. Maintaining constant speed over time (zero acceleration phase) will allow for the influence of added mass to be removed, making estimation of the drag terms easier. Maintaining constant speed can, of course, be done with the PMM and the chain and pulley system.

3.4 Computational Method and Numerical ROV Modeling

As computers and computational software get relatively cheaper over time, it becomes more convenient and expedient to use such a tool to determine the hydrodynamic forces and parameters. A typical approach is to use a computer-aided design (CAD) tool (e.g. SolidWorks™) to "build" the model of the vehicle, and with the knowledge of geometry, the mechanical design (with constraints) and materials used for various mechanical components, the rigid body mass matrix M_{RB} with the masses and the mass moments of inertia, and the location of the center of mass can be determined.

CAD tools integrated with packages like MULTISURF™ or Gambit™ can create surface geometry data for use with a software WAMIT™ (Wave Analysis, MIT) or MCC (Marine Craft Characteristics) to produce added mass matrix M_A. To determine the damping forces, finite-element computational fluid dynamics software like ANSYS CFX™ can be used. It can be viewed as a software version of a water tunnel, where flow past the stationary ROV is computed.

Comparison of the performance of some of this software can be found in [9]. The authors of [9] found no significant difference in the M_A matrix with WAMIT™ and MCC software, but for drag forces in rotational motions, they noted that ANSYS CFX™ computations require re-meshing and is expensive, and the results were considered inappropriate compared with the use of Siemens Simcenter STAR-CMM+™. The Siemens-integrated STAR-CMM+™ is a multi-physics solution. The successful determination of drag forces using STAR-CMM+™ for open frame ROV [10, 11] was reported.

3.5 Validation by Simulation or Experiments

If the suite of computer software for marine vessels is available, it will be a reasonable starting platform—faster setup and comparatively cheaper. And with computer modeling, it is often necessary to check the model creation approach that produces results comparable with that predicted by theories; for example, with flow past a sphere or a cylinder the theoretical results and the computational model results should be sufficiently close to each other. The same approach is used to create the computer model of the vehicle, and both the model and the results must then be validated against experimental results in the tank [7, 9] or as reported by some authors by model simulation [6, 10, 11].

3.6 Proposed Benchmark ROV Model

As part of the effort in this book to undertake the feasibility study and development of a benchmark ROV model, we have produced a comprehensive dynamic model for the ROV. ROV is inherently highly coupled, nonlinear, and uncertain due to the hydrodynamic forces and external environmental forces. Since there are no standard ROV model available for control system design and analysis, the benchmark ROV is proposed.

3.6.1 Numerical ROV Modeling

In the proposed ROV modeling approach [11], Fig. 3.3 shows the overall approach of numerical modeling to experimental validation of the benchmark ROV model. The computer-aided design (CAD) model of an ROV will be created using software such as SolidWorks™ and MULTISURF™. With the CAD model, the hydrodynamic parameters like hydrodynamic damping and added mass coefficients for the nonlinear ROV model are determined using WAMIT™ and MATLAB™. The obtained ROV model is later validated experimentally.

Figure 3.3 shows the geometrical model of the ROV modeled by SolidWorks™. The ROV model is actuated by six DC brushless thrusters (T1–T6) for the surge, sway, heave, roll, pitch, and yaw. The mechanical properties such as mass, moment

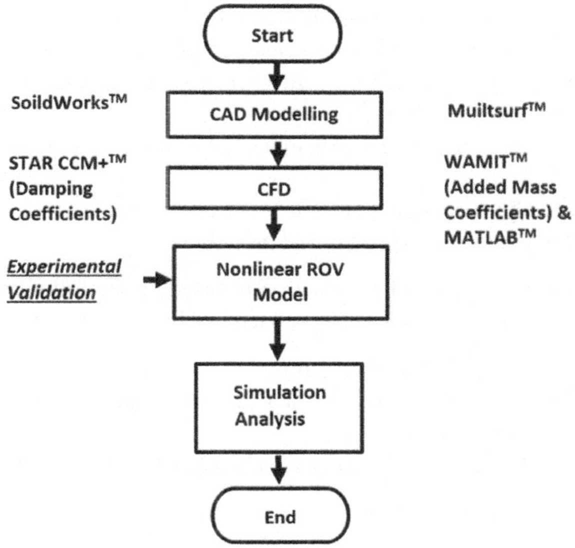

Fig. 3.3 Overall flowchart of proposed systematic computation of ROV model [11]

of inertia, the center of gravity, and buoyancy of the ROV are obtained. In Fig. 3.4, the ROV has a mass of approximately 75 kg in the air with an overall dimension of 1455 mm (length) × 950 mm (width) × 400 mm (height).

The following sensor and actuators [10] are connected to a target PC using Diamond Systems ATHENA III PC104 operating in Ubuntu platform. The main components used in the ROV are as follows.

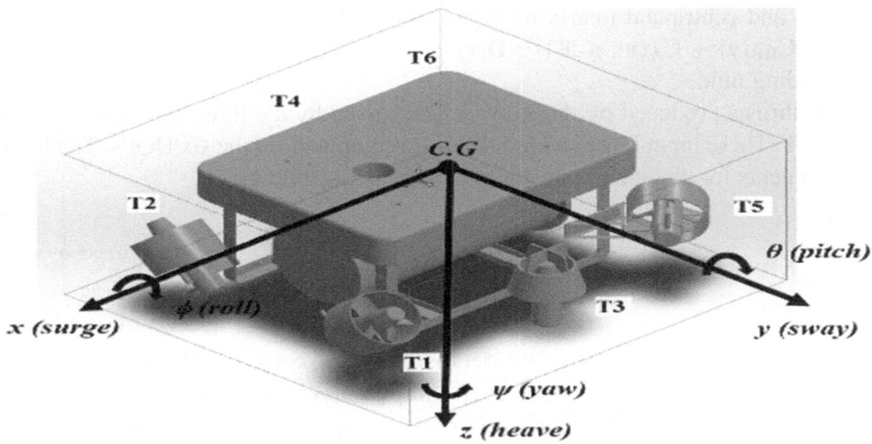

Fig. 3.4 CAD SolidWorks™ model for ROV and its body-fixed coordinate system [11]

(1) NavQuest Doppler Velocity Log (DVL) to measure the velocity of ROV;
(2) Druck pressure sensor or altimeter is used to measure the depth of the ROV;
(3) XSENS inertia measurement unit (IMU) is used to measure the positions and angular positions of the ROV;
(4) Two USB cameras for underwater image capturing; USB Joystick for remote control;
(5) Six DC Tecnadyne brushless electric thrusters connected to an ISO-6 thruster driver card to control the speed of the propellers.

The derivation of ROV dynamic equation is based on the following assumptions to simplify the ROV model. The same approach in Chap. 1 is used to model the ROV.

- Operating at slow speed (less than 1 m/s);
- Rigid body and fully submerged in water (no wave and current disturbance);
- Neutrally buoyant by design;
- Tether (or umbilical cable) dynamics attached to ROV is not considered.

The following equation of motions for the ROV is formed:

$$\mathbf{M}\dot{\boldsymbol{\nu}} + \mathbf{C}(\boldsymbol{\nu})\boldsymbol{\nu} + \mathbf{D}(\boldsymbol{\nu}) + \mathbf{G_f}(\boldsymbol{\eta}) = \boldsymbol{\tau} \tag{3.7}$$

where $\boldsymbol{\nu} = \begin{bmatrix} u\ v\ w\ p\ q\ r \end{bmatrix}^{\mathrm{T}}$ is the body-fixed velocity vector, and $\boldsymbol{\eta} = \begin{bmatrix} x\ y\ z\ \phi\ \theta\ \psi \end{bmatrix}^{\mathrm{T}}$ is the earth-fixed vector comprising the position vector. $\mathbf{M} = \mathbf{M_{RB}} + \mathbf{M_A} \in \Re^{6\times 6}$ is the inertia matrix for rigid body and added mass, respectively. The gravitational and buoyancy vector acting on ROV in water is denoted by $\mathbf{G_f}(\boldsymbol{\eta})$ [1]. The ROV is neutrally buoyant (W = B) and the X–Y coordinates of the C.B. coincide with the X–Y coordinate of the CG (by placing additional mass on ROV). Since $(z_G - z_B) = -0.397$ m, the resulting $\mathbf{G_f}(\boldsymbol{\eta})$ can be expressed as $\mathbf{G_f}(\boldsymbol{\eta}) = [0\,0\,0\ \ -0.397\mathrm{W}\cos\theta\sin\phi\ \ -0.397\mathrm{W}\sin\theta\ \ 0]^{\mathrm{T}}$. The Coriolis and centripetal matrix for rigid body and added mass can be defined as: $\mathbf{C}(\boldsymbol{\nu}) = \mathbf{C_{RB}}(\boldsymbol{\nu}) + \mathbf{C_A}(\boldsymbol{\nu}) \in \Re^{6\times 6}$; $\mathbf{D}(\boldsymbol{\nu}) \in \Re^{6\times 6}$ is the damping matrix due to the surrounding fluid.

The thrusts produced by the ROV are expressed by the thrusters' configuration matrix, **T**. These input force and moments are computed for the six DOFs. The force and moments for the open-hoop configuration are as follows:

$$\begin{aligned}
\mathrm{u}_3\cos\alpha + \mathrm{u}_4\cos\alpha &= \tau_x \\
-\mathrm{u}_1\cos\beta + \mathrm{u}_2\cos\beta + \mathrm{u}_3\sin\alpha - \mathrm{u}_4\sin\alpha &= \tau_y \\
\mathrm{u}_1\cos\beta + \mathrm{u}_2\cos\beta + \mathrm{u}_5 + \mathrm{u}_6 &= \tau_z \\
0.155\mathrm{u}_1\cos\beta - 0.155\mathrm{u}_2\cos\beta - 0.275\mathrm{u}_5 + 0.275\mathrm{u}_6 &= \tau_\varphi \\
0.3945\mathrm{u}_1\cos\beta + 0.3945\mathrm{u}_2\cos\beta + 0.4305\mathrm{u}_3\cos\beta + 0.4305\mathrm{u}_4\cos\beta - 0.0355\mathrm{u}_5 - 0.0355\mathrm{u}_6 &= \tau_\theta \\
-0.3945\mathrm{u}_1\sin\beta + 0.3945\mathrm{u}_2\sin\beta - 0.6605\mathrm{u}_3\sin\beta + 0.6605\mathrm{u}_4\sin\beta &= \tau_\psi
\end{aligned} \tag{3.8}$$

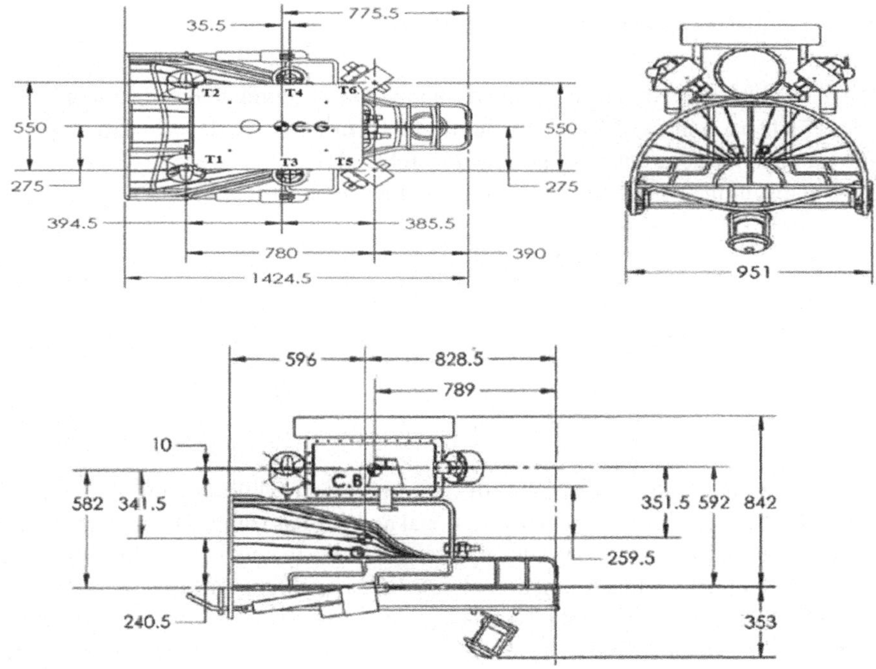

Fig. 3.5 Six thrusters (namely T1, T2, T3, T4, T5, and T6) position on ROV platform

where $u_{min} \leq u_i \leq u_{max}$, $i = 1 \ldots 6$ are the limits of thrust output and $\boldsymbol{\tau} = [\tau_x\ \tau_y\ \tau_z\ \tau_\phi\ \tau_\theta\ \tau_\psi]^T$ is the thrust vector generated. Here α is the orientation angle for T3 and T4, while β is the orientation angle for T1 and T2. Similar equations can be established for the closed-hoop configuration using different values in Fig. 3.5.

The thruster configuration matrix **T** based on the layout of the thrusters in ROV platform is defined as follows:

$$\mathbf{T} = \begin{bmatrix} 0 & 0 & 0 & 0 & \cos\alpha & \cos\alpha \\ -\cos\beta & \cos\beta & 0 & 0 & \sin\alpha & -\sin\alpha \\ \cos\beta & \cos\beta & 1 & 1 & 0 & 0 \\ 0.155\cos\beta & -0.155\cos\beta & -0.275 & 0.275 & 0 & 0 \\ 0.394\cos\beta & 0.394\cos\beta & -0.035 & -0.035 & 0.430\cos\beta & 0.430\cos\beta \\ -0.394\sin\beta & 0.394\sin\beta & 0 & 0 & -0.660\sin\beta & 0.660\sin\beta \end{bmatrix} \tag{3.8}$$

where $\alpha = 45^{\circ}$ is the inclination angle for T5 and T6, while $\beta = 45^{\circ}$ is the orientation angle for T1 and T2. Note that T3 and T4 are the two vertical thrusters shown in Fig. 3.4. The computation of hydrodynamic added mass and damping coefficients are discussed in the following sections.

3.6.2 Hydrodynamic Damping Model

The hydrodynamic damping coefficients of the ROV are computed in the report [11]. The vehicle has a mass of 75 kg, a volume of 0.05 m^3, and a surface area of 4.75 m^2. The moment of inertia of ROV is shown in Table 3.1.

As a result, the rigid-body ROV mass inertia can be written as:

$$\mathbf{M_{RB}} = \begin{bmatrix} 75.00 & 0 & 0 & 0 & 0 & 0 \\ 0 & 75.00 & 0 & 0 & 0 & 0 \\ 0 & 0 & 75.00 & 0 & 0 & 0 \\ 0 & 0 & 0 & 2.51 & 0 & 0 \\ 0 & 0 & 0 & 0 & 3.38 & 0.01 \\ 0 & 0 & 0 & 0 & 0.01 & 1.73 \end{bmatrix} \quad (3.9)$$

A submerged body experiences lift and drag effect while moving through the fluid. This drag component includes frictional and pressure drag. The frictional drag due to the boundary layers depends on the surface area in contact with the fluid. The damping function is a linear function of velocity, a quadratic function of velocity, a sum of both linear and quadratic terms (as used in this paper) and with higher-order forms. The steady drag force experienced by the vehicle in its reference state is well known to depend on the square of the velocity and a coefficient that depends on Reynolds number (for a body sufficiently submerged in a fluid). The variation of that force on the body experienced during small perturbations to that motion has repeatedly found to be better modeled by a linear function of velocity. The linear coefficients are therefore adequate to represent the strength and moments due to inviscid part of the flow for a low-speed ROV.

The ROV hydrodynamic damping matrix **D** can be further simplified. The off-diagonal elements [1] in the hydrodynamic damping matrix **D(v)** are small compared to those diagonal elements on the underwater vehicle. Therefore, **D(v)** becomes a diagonal matrix: $\mathrm{diag}[\{X_u, Y_v, Z_w, K_p, M_q, N_r\}]$. A turbulence model with the unsteady three-dimensional flow was built for the Reynolds number flow condition greater than 1.0×10^6. The shear–stress transport (SST) model in CFD used STAR-CCM + software. The k-ω SST model is one of the two common models for predicting the flow separation under an adverse pressure gradient. It provides a highly accurate prediction of the amount of flow separation under adverse eddy-viscosity. To take advantage of the SST model, the boundary layer should be resolved with at

Table 3.1 Moment of inertia properties of ROV [11]

Moment of inertia (kg m^2)					
I_{xx}	2.51	I_{xy}	0	I_{xz}	0
I_{yx}	0	I_{yy}	3.38	I_{yz}	0.01
I_{zx}	0	I_{zy}	0.01	I_{zz}	1.73

least 10 mesh nodes. This is done by inspecting the y + value on the surface of the ROV that must be around one.

The movement in the fluid domain is expected to be turbulent and isothermal. The temperature is fixed at 20 °C and the water is modeled as an incompressible fluid. It is impractical to set the fluid domain to be infinitely large to analyze damping force acting on ROV in CFD. As a result, a dimension of approximately 15 times greater than the dimension of the ROV (see Fig. 3.6) to ensure the accuracy [5] of

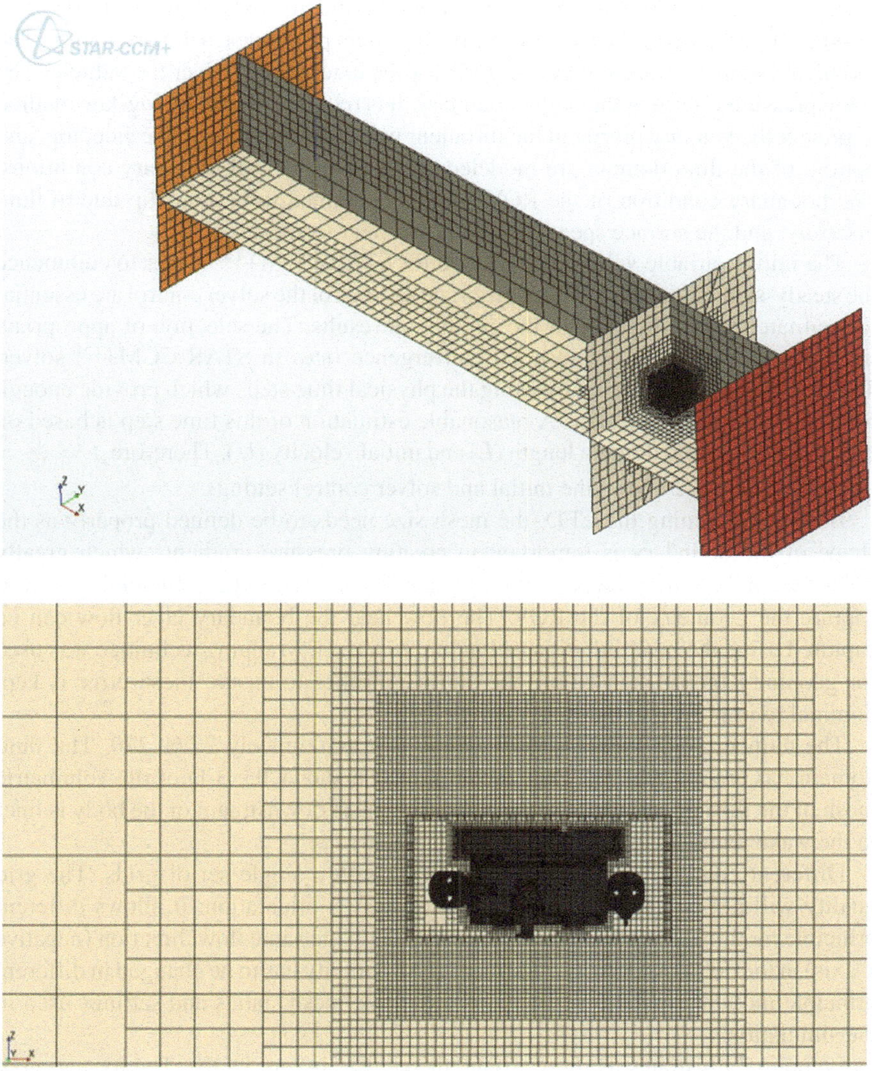

Fig. 3.6 3-D view of volumetric meshing for flow domain (top) and close view of ROV (bottom) [11]

Table 3.2 Initial condition and solver control settings [11]

Setting	Value
Initial conditions > Turbulence intensity	0.1%
Stopping criteria > Maximum steps	60
Physical timescale control	5 [s]

the actual flow domain is used. The properties of the flow domain are required to be defined to ensure the domain has high fidelity. The turbulence properties are difficult to define without turbulent kinetic energy and dissipation data at the inlet flow. The turbulent intensity was used to describe the turbulent properties in the simulation. The turbulent intensity is selected to be 0.1% for the external flow over the vehicle. The static pressure is used at the outlet boundary. It is required to use the log-law method to predict the velocity profile in the turbulent wall-bounded flows. The side, top, and bottom of the fluid domain are modeled using free slip wall boundary conditions. The boundary condition of the ROV surface is defined to be non-slip due to fluid viscosity, and the surface speed of the ROV is near to zero.

The initial variable values are required for STAR-CCM+^{TM} solver to commence the steady-state calculation. The appropriate settings of the solver control are essential to facilitate the convergence of the simulation results. The selection of appropriate time step size helps to obtain good convergence rates in STAR-CCM+^{TM} solver. The convergence is achieved by using the physical time steps which provide enough relaxation of the nonlinearity. A reasonable estimation of this time step is based on one-third of the fluid domain length (L) and initial velocity (U). Therefore, $t = \frac{L}{3U} = \frac{15}{3*1} = 5s$. Table 3.2 shows the initial and solver control settings.

Before performing the CFD, the mesh size needs to be defined properly as the shape of the boundary is important in creating pressure gradients which greatly influence the boundary layer. The mesh size is preferred to be sufficiently small to capture the geometry of the ROV. The flow near the boundary layer flow can be captured using the layer inflation technique. A surface wrapping technique was used for geometry preparation before the surface meshes to ensure mesh error is kept minimal to improve the meshing quality.

The number of elements in the volume is approximately 2,069,270. The fluid domain has around 546,712 elements. Figure 3.6 shows the 3-D of the volumetric mesh of the flow domain around the ROV. The mesh downstream of the body is finer in the wake region than near the domain boundary.

Different configurations can be computed with a single set of grids. The grid quality will not be affected by changing of ROV's orientation. It allows different principle motions of the ROV to be simulated with the same flow direction (negative x-axis) in the domain. However, the direction of gravity has to be changed in different principle motion simulation. Table 3.3 shows the mesh statics and settings used in the simulation.

Figure 3.7 illustrates the flow around the vehicle in four DOFs. The formation of the wake at the rear creates a low-pressure region. The high-pressure region at the front resists the motion of the vehicle. As observed, there is certainly flow separation

Table 3.3 Mesh settings [11]

Mesh setting	Value
No. of mesh elements	Around 2.6 million
Mesh spacing	0.0005 [m^2]
Boundary growth rate (i.e. rate at which the boundary layer thickness grows)	Low
Volumetric mesh type	Polyhedral, Trimmer

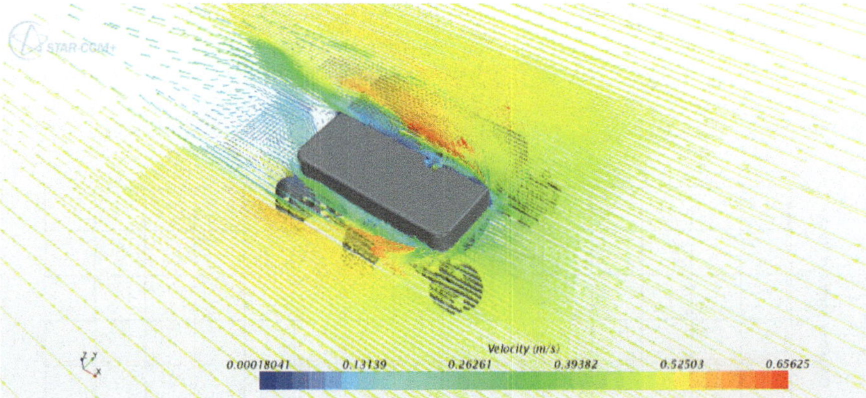

Fig. 3.7 Vector plot of ROV in STAR-CCM+^{TM} [11]

on the ROV. Figure 3.8 illustrates the drag force exerting on the ROV and its corresponding drag coefficient. It shows that the drag forces converge to a steady-state value at around 40 iterations.

As observed in Fig. 3.7, the wakes are formed at the rear of the ROV body when the turbulent flow past the vehicle. The wakes represent the increase in the vortices of the flow separation. The nonlinear damping effects on the ROV are caused by this flow separation phenomenon in the turbulent model. The hydrodynamic damping forces were then determined by integrating the pressure on the ROV surface. The three translation motions, namely surge, sway, and heave, are plotted against the velocity in Fig. 3.9. As observed from the plots, the vehicle has the largest drag in the heave direction due to its largest frontal area normal to the flow direction of about 0.6 m^2. The drag force in sway is slightly larger than the drag in surge direction due to its small frontal area. The damping coefficient for different directions is obtained using a second-order polynomial fit as shown in Figs. 3.9 and 3.10.

Table 3.4 shows the drag coefficient results of ROV in its four principle motions. The flow in the domain was simulated at different speeds using STAR-CCM+. Table 3.5 tabulates the drag moment versus angular velocity in the yaw direction.

Figure 3.10 depicts the drag moment as a function of square angular velocity. The linear damping in yaw direction is ignored as only quadratic damping is present. The

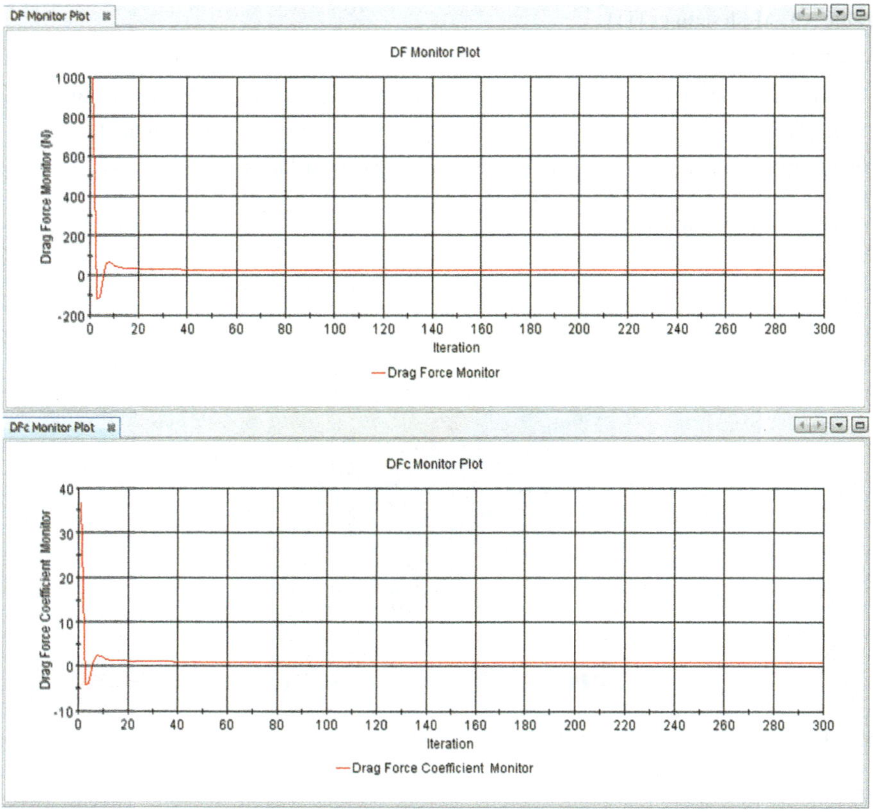

Fig. 3.8 Iteration history of drag force and its coefficient [11]

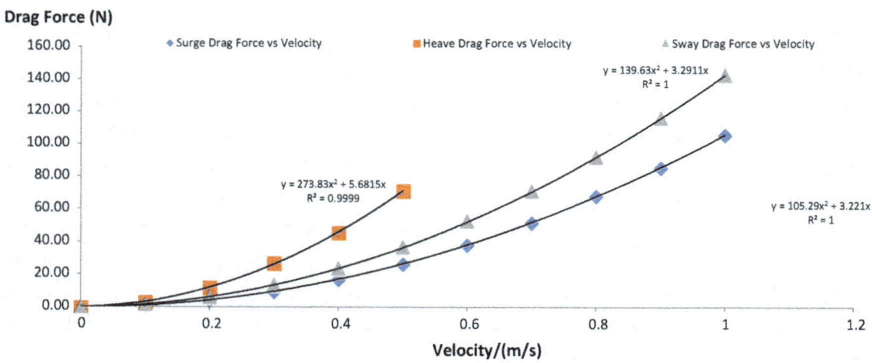

Fig. 3.9 Drag force as the function of velocity in surge, sway, and heave [11]

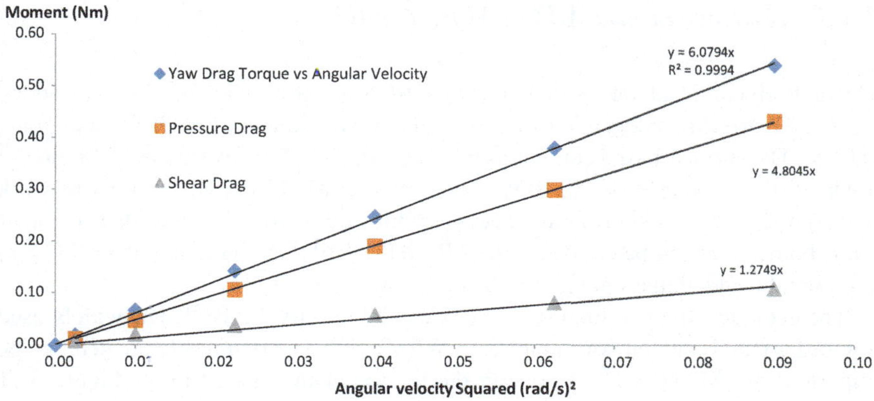

Fig. 3.10 Drag torque as a function of velocity in yaw [11]

Table 3.4 Damping coefficients of ROV in four principle directions [11]

Damping coefficients	Surge		Sway		Heave		Yaw	
	K_L	K_Q	K_L	K_Q	K_L	K_Q	K_L	K_Q
Values	3.221	105.3	3.291	139.6	5.682	273.8	0	6.079

Table 3.5 Drag moment versus angular velocity in yaw direction [11]

Angular velocity (rad/s)	Yaw moment (Nm)	Yaw moment coefficient
0.100	0.067	0.204
0.200	0.248	0.188
0.300	0.542	0.183

resulted hydrodynamic damping for the ROV is thus written as:

$$\mathbf{D_L} = -\mathrm{diag}\{[\,3.221\ 3.291\ 5.682\ 0.\ 0\ 0\,]\} \tag{3.10}$$

$$\mathbf{D_Q} = -\mathrm{diag}\ \{[\,105.300\ 139.600\ 273.800\ 0\ 0\ 6.079\,]\} \tag{3.11}$$

In summary, the results show that the lowest damping occurs in surge direction while the heave motion has the largest drag force due to its larger surface area in contact with water. The values of the linear damping coefficients are smaller than the nonlinear damping terms due to its square velocity term.

3.6.3 Hydrodynamic Added Mass Model

The hydrodynamic added coefficients of the ROV are now analyzed [11]. The added mass and inertia are independent of the circular wave frequency for a fully submerged vehicle. The surface-based computer-aided design (CAD) software MULTISURF™ modeled the geometry of the ROV. The software MULTISURF™ aims to work with WAMIT™ for exporting the necessary files for analysis. The ROV is made of multi-bodies that can be created using MULTISURF™ and later use WAMIT™ to compute the added mass matrix as shown below.

The geometry file was imported to WAMIT™ by MATLAB™. The widely used low-order panel method was adopted in WAMIT™. The output file created can be imported into MATLAB™ to obtain the hydrodynamic added mass. Figure 3.11 illustrates the procedure for computing the hydrodynamic added mass coefficients.

Table 3.6 shows the descriptions of each file (see Fig. 3.11) used in determining the added mass coefficients. All the necessary parameters required are shown in the following input files. Before using the WAMIT™ to test the ROV, a study on the

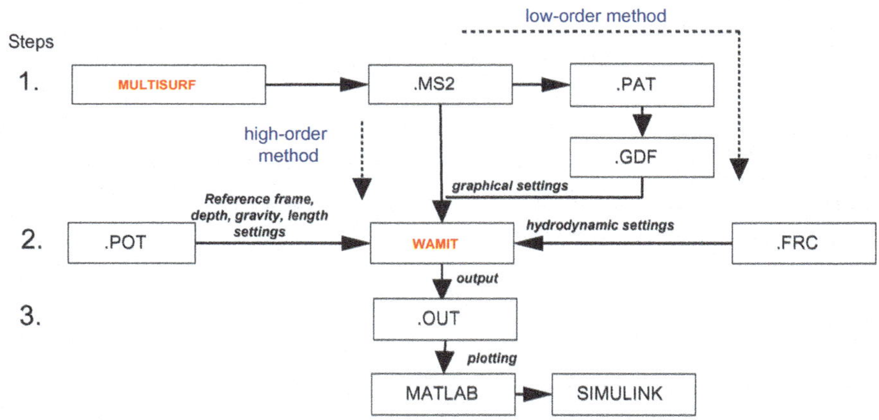

Fig. 3.11 Program flowchart for computing added mass coefficients [11]

Table 3.6 Description of records used in WAMIT™ solver

File extension	Description
.GDF	Geometry data file used to define the geometry in panel form
.POT	Potential control file used to define input parameters in POTEN (length, gravity)
.FRC	Force control file used to define input parameters in FORCE (desired hydrodynamic parameters)
.OUT	Output from WAMIT
readAM.m	MATLAB m-file used to read added mass matrix

empirical results of a 2 m diameter sphere (see Fig. 3.12) was used to verify the program setup and parameters. In Table 3.7, the theoretically added mass of a sphere is $A = 2/3\pi\rho r^3$ for the three translational motions: surge, sway, and heave. The added mass of the sphere can be written as $A = 2/3\pi r^3$ (after normalizing with density). Table 8 shows that the results of low-order method of the sphere are quite close to the theoretical value of 2.094.

Additionally, it is required to define the depth of the submerged body in WAMIT™. The impact of different water depth has been identified by using the same sphere. As shown in Table 3.8, the results of the added mass converged at approximately 10 m. As a result, the computation will be performed at 10 m.

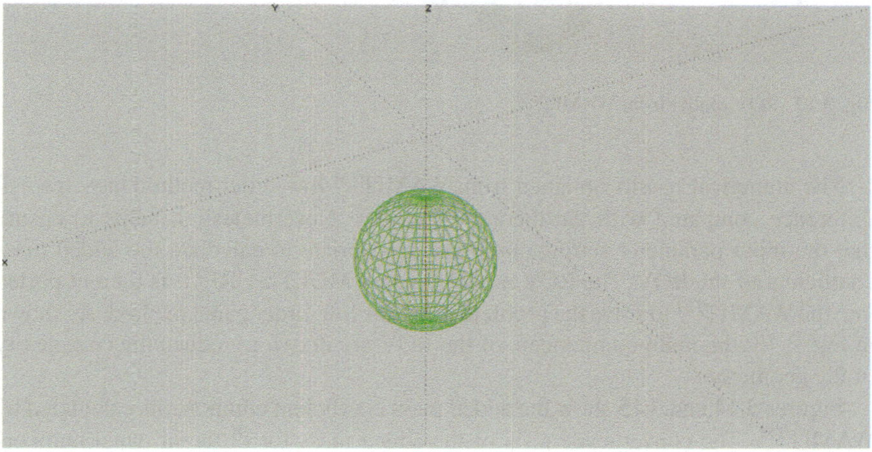

Fig. 3.12 Finite surface panels generation of sphere using MULTISURF™ [11]

Table 3.7 Low-order method for sphere [11]

Panel number	Numerical results			Theoretical results		
	Surge (m^3)	Sway (m^3)	Heave (m^3)	Surge (m^3)	Sway (m^3)	Heave (m^3)
256	2.085	2.085	2.073	2.094	2.094	2.094
512	2.084	2.084	2.087			
1024	2.083	2.083	2.091			
Errors				−0.5%	−0.5%	−0.1%

Table 3.8 Added mass of sphere at different depth [11]

Depth (m)	Added mass (kg)
0	2.5910
1	2.1419
10	2.0838
100	2.0835

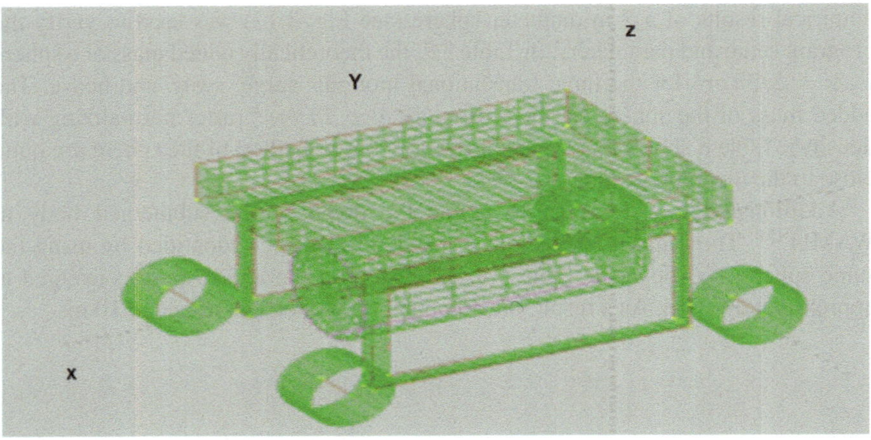

Fig. 3.13 ROV model from WAMIT^{TM}

The numerical results obtained from WAMIT^{TM} low-order method have a small difference compared with the theoretical results. Nevertheless, it helps to ensure that the input parameter settings used are appropriate to calculate the added mass coefficient of the ROV. The ROV modeled using MULTISURF^{TM} is then imported into the WAMIT^{TM} to solve the problem using the low-order panel method. As shown in Fig. 3.13, the main components of the ROV are drawn to reduce the complexity of the geometry.

Figures 3.14 and 3.15 show the added mass coefficient components calculated by WAMIT^{TM}. The convergence tests of the added mass for different panel numbers converged approximately to 3000–4000 panels. It shows that the minimum required panel numbers should be at least 3000.

The off-diagonal terms of the added mass are less than the diagonal components as the ROV has near to three planes of symmetry. Hence, the added mass matrix obtained is further simplified.

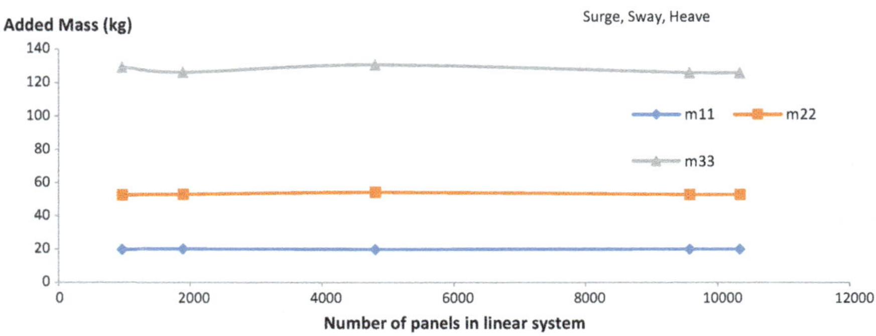

Fig. 3.14 Convergence plot for surge, sway, and heave direction [11]

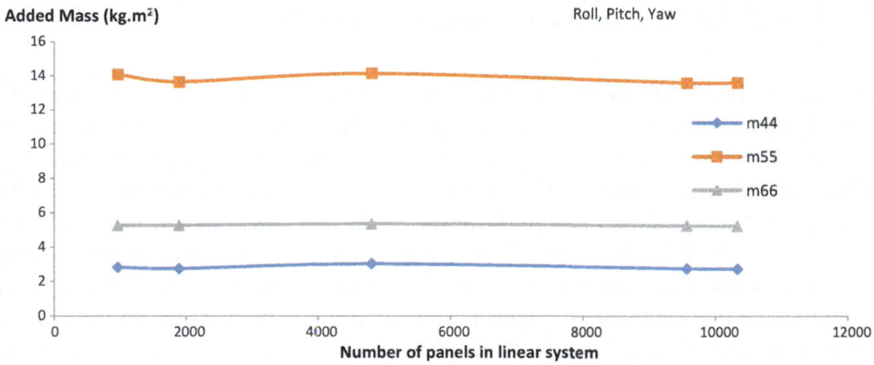

Fig. 3.15 Convergence plot for roll, pitch, and yaw direction [11]

$$\mathbf{M_A} = -\begin{bmatrix} 20.392 & 0 & 0 & 0 & 0 & 0 \\ 0 & 53.435 & 0 & 0 & 0 & 0 \\ 0 & 0 & 126.144 & 0 & 0 & 0 \\ 0 & 0 & 0 & 2.802 & 0 & 0 \\ 0 & 0 & 0 & 0 & 13.703 & 0 \\ 0 & 0 & 0 & 0 & 0 & 5.263 \end{bmatrix} \tag{3.12}$$

The added mass matrix $\mathbf{M_A}$ of the ROV must be positive. The following relations are observed from the matrix: $m_{11} < m_{22} < m_{33}$. A smaller added mass in surge direction, that is, m_{11} has been observed. It is due to the ROV has the smallest projection area in surge direction. The corresponding Coriolis and centripetal added mass matrix can be rewritten as:

$$\mathbf{C_A}(\boldsymbol{\nu}) = \begin{bmatrix} 0 & 0 & 0 & 0 & -126.144w & 53.435v \\ 0 & 0 & 0 & 126.144w & 0 & -20.392u \\ 0 & 0 & 0 & -53.435v & 20.392u & 0 \\ 0 & -126.144w & 53.435v & 0 & -5.263r & 13.703q \\ 126.144w & 0 & -20.392u & 5.263r & 0 & -2.802p \\ -53.435v & 20.392u & 0 & -13.703q & 2.802p & 0 \end{bmatrix} \tag{3.13}$$

In summary, the added mass coefficients of the ROV are around 20 kg in surge direction, 53 kg in sway direction, followed by 126 kg in heave direction. The ROV has the larger added mass in heave, followed by the sway and surge direction. In summary, the computations of the hydrodynamic damping and added mass were performed. The proposed numerical simulation can determine the hydrodynamic parameters.

3.6.4 *Validation Using Experimental Results*

The results obtained from the simulations were validated with the experimental data [10, 11] in a water tank. The heave and yaw motions were only validated due to the limited reliable sensor installed on the ROV. A depth sensor, Doppler velocity log (DVL), and inertial measurement unit (IMU) were jointly used to measure the depth, velocity, and acceleration of the ROV, respectively. The water tank dimensions are: 10 m (L) × 4 m (W) × 1.8 m (D). It is equipped with an overhead crane to load and unload the ROV. The sample rate of 100 Hz (or sample period of 0.01 s) was used to sample the raw data from the sensor. The raw data are plotted against the sample count using a MATLAB^{TM} script. Note that one sample count is equivalent to 0.01 s.

3.6.4.1 Heave Model Identification

During the depth validation, the prototype ROV (see Fig. 3.16) was maintained at a fixed heading angle before submerging to a depth of 1 m, as shown in Fig. 3.17. The ROV was commanded to move vertically into the water with little acceleration measured by IMU. The dead-reckoning data for acceleration in Z-direction is not

Fig. 3.16 Image of ROV (top) and motion pictures (bottom) captured during heave test in water

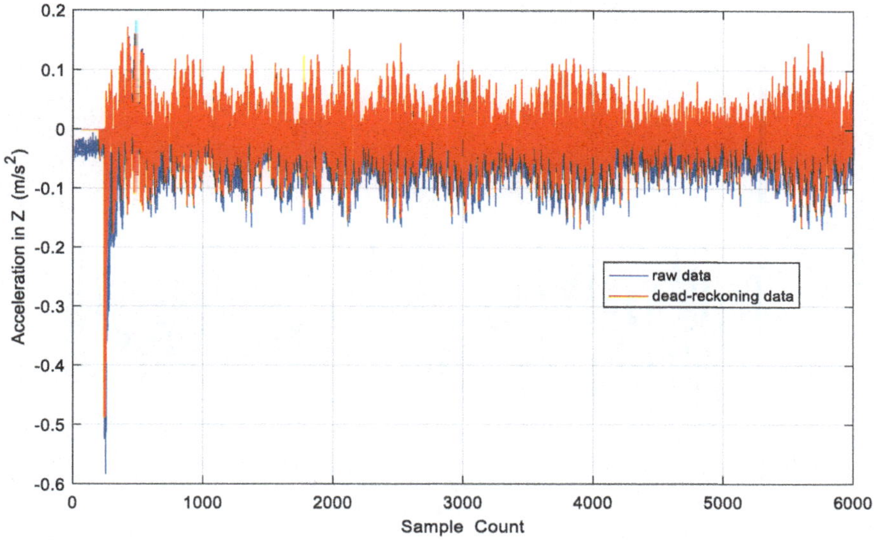

Fig. 3.17 Acceleration in heave direction (1 sample count = 0.01 s)

zero, as shown in Fig. 3.17. The ROV is not moving at a constant heave rate as observed in the heave velocity from the DVL in Fig. 3.18. The thrust outputs from T3 and T4 (for heave direction) are approximately 10 N (see Fig. 3.19) to allow the ROV to reach a depth of 1 m, as depicted in Fig. 3.20.

The added mass $Z_{\dot{w}}$ and damping coefficients (linear, Z_w) due to the hydrodynamic force are defined in a body-fixed frame with m as the body mass of the ROV. The heave model can be simplified:

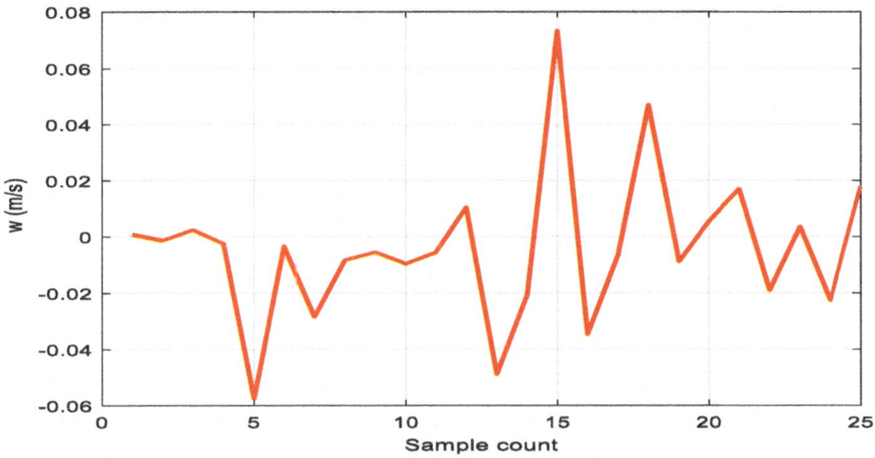

Fig. 3.18 Heave velocity response (1 sample count = 0.01 s)

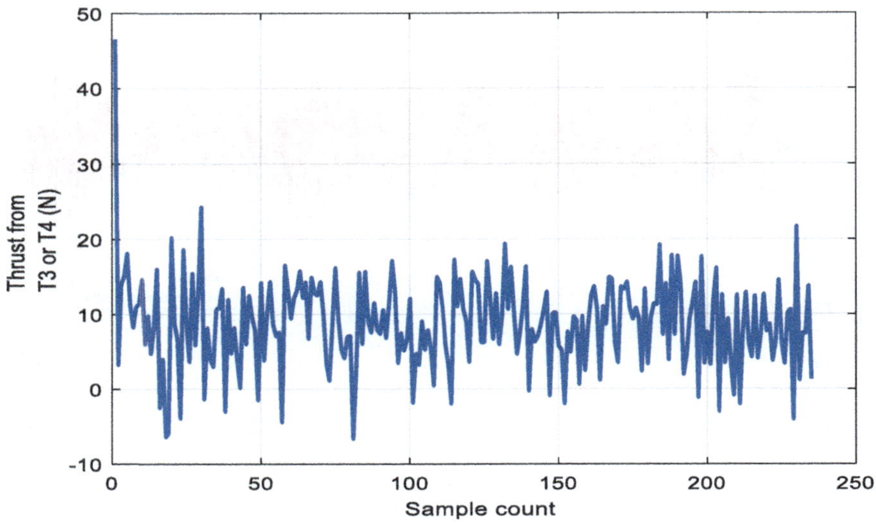

Fig. 3.19 Thrust output from thruster T3 or T4 (1 sample count = 0.01 s)

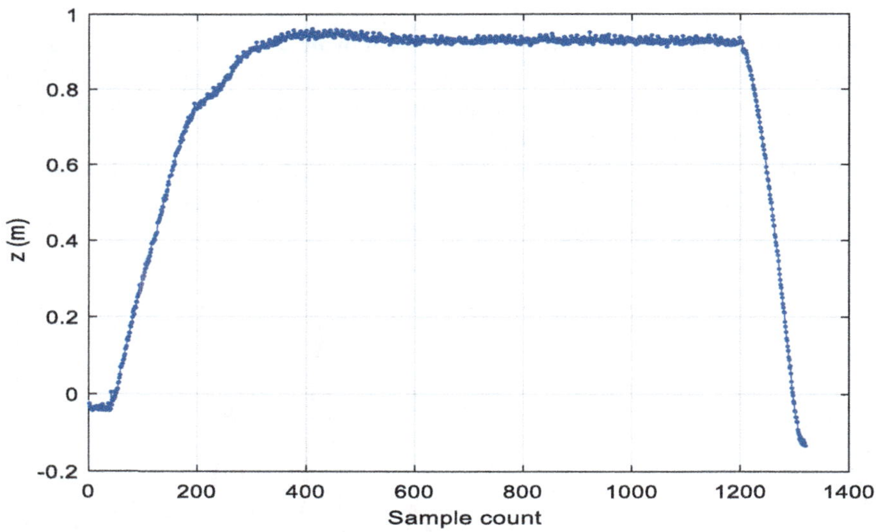

Fig. 3.20 Depth response (1 sample count = 0.01 s)

$$(m + Z_{\dot{w}})\dot{w} + Z_w w = Z \tag{3.14}$$

The ROV is designed to be neutrally buoyant, and the centripetal and Coriolis terms are not present in the equation as the ROV was commanded to move in heave direction only. Rearranging the preceding equation gives:

$$\dot{w} = \frac{Z}{(m + Z_{\dot{w}})} - \frac{Z_w}{(m + Z_{\dot{w}})} w \tag{3.15}$$

In matrix form,

$$\overbrace{\dot{w}}^{\varphi} = \overbrace{\begin{bmatrix} Z & -w \end{bmatrix}}^{\mathbf{H}} \overbrace{\begin{bmatrix} \alpha \\ \beta \end{bmatrix}}^{\hat{\theta}} \tag{3.16}$$

where

$$\alpha = \frac{1}{(m + Z_{\dot{w}})}; \beta = -\frac{Z_w}{(m + Z_{\dot{w}})} \tag{3.17}$$

The following least-square method is used to obtain the estimated values.

$$\underbrace{\begin{bmatrix} \dot{w}_1 \\ \dot{w}_2 \\ \vdots \end{bmatrix}}_{\varphi} = \underbrace{\begin{bmatrix} Z_1 & -w_1 \\ Z_2 & -w_2 \\ \vdots & \vdots \end{bmatrix}}_{\mathbf{H}} \cdot \underbrace{\begin{bmatrix} \alpha \\ \beta \end{bmatrix}}_{\theta} + \text{error} \tag{3.18}$$

where the subscript i represents the number of samples collected from the experiment. The standard least-square solution is given by the Moore–Penrose pseudoinverse:

$$\hat{\theta}_{\mathbf{LS}} = (\mathbf{H}^{\mathbf{T}}\mathbf{H})^{-1} \cdot \mathbf{H}^{\mathbf{T}} \cdot \varphi \tag{3.19}$$

where the standard deviation, $\text{cov}(\hat{\theta}_{\mathbf{LS}}) = \sigma^2(\mathbf{H}^{\mathbf{T}}\mathbf{H})^{-1}$ and note that $\mathbf{H}$ has to be a full rank matrix. The recursive least square (RLS) is used to compute the parameters $\hat{\theta}_{\mathbf{RLS}}(t)$ in (3.19). The parameters update $\hat{\theta}_{\mathbf{RLS}}(t)$ includes a correction term to the previous estimate $\hat{\theta}_{\mathbf{RLS}}(t-1)$ as seen in (3.20). The typical RLS algorithm comprises the following recursive equations computed in sequence. For example, $\hat{\theta}_{\mathbf{RLS}}(t-1)$ in (3.20) requires the computation of (3.21)–(3.23).

$$\hat{\theta}_{\mathbf{RLS}}(t) = \hat{\theta}_{\mathbf{RLS}}(t-1) + \mathbf{K}(t)\varepsilon(t) \tag{3.20}$$

$$\boldsymbol{\varepsilon}(t) = \dot{\mathbf{w}}(t) - \hat{\theta}^{\mathbf{T}}_{\mathbf{RLS}}(t-1)\mathbf{H}(t) \tag{3.21}$$

$$\mathbf{K}(t) = \frac{\lambda^{-1}\mathbf{P}(t-1)\mathbf{H}(\text{t})}{1 + \lambda^{-1}\mathbf{H}^{\text{T}}\mathbf{P}(t-1)\mathbf{H}(t)} \tag{3.22}$$

$$\mathbf{P}(t) = \lambda^{-1}\mathbf{P}(t-1) - \lambda^{-1}\mathbf{K}^{\text{T}}(t)\mathbf{H}(t)\mathbf{P}(t-1) \tag{3.23}$$

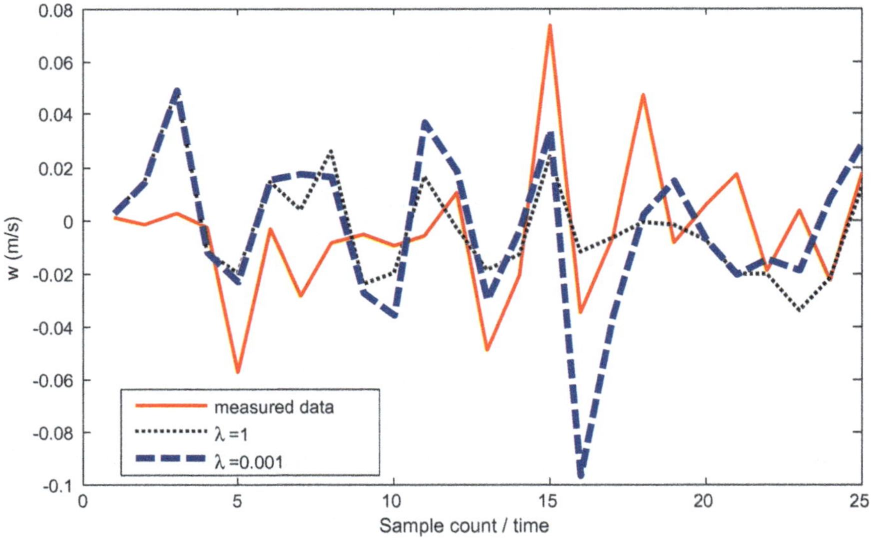

Fig. 3.21 Comparison of heave velocity response using RLS

Table 3.9 Comparisons between simulation and experimental results in heave direction

Descriptions	Simulation	Experiment (Recursive Least Square)	Error (%)
Added mass (kg)	126.14	136.40	8
Linear damping coefficient	5.6800	4.0874	28
Quadratic damping coefficient	273.805	–	–

The above equations need an initial value for $\hat{\theta}_{\mathbf{RLS}}(t)$ and the error covariance matrix $\mathbf{P}$. The initial value $\hat{\theta}_{\mathbf{RLS}}(t)$ is set as 0, and $\mathbf{P}$ is $100\mathbf{I}_2$ where $\mathbf{I}_2$ is the identity matrix of dimension two. The forgetting factor that specifies how fast the RLS forgets the past sample information is set to 0.001. For example, using $\lambda = 1$, it specifies an infinite memory. By comparing the RLS with a different forgetting factor, the following plot in Fig. 3.21 is obtained. In Fig. 3.21, the RLS with a lower forgetting factor has a closer behavior to the measured data. In general, it can be observed that the differences in the performance are quite large for the damping in the heave direction as seen in Table 3.9. It may be because the effect of the rotating propeller on the thrusters is not included during the simulation.

3.6.4.2 Yaw Model Identification

A simplified description of the yaw model of the ROV is defined as

$$N_{\dot{r}}\dot{r} + N_{|r|r}|r|r + N_r r = N \tag{3.24}$$

$$\dot{\psi} = r \tag{3.25}$$

where $N_{\dot{r}}$ is the added inertia ($= I_r + N_{\dot{r}}$) in the yaw direction. N_r is the linear component of drag force in the yaw direction and $N_{|r|r}$ is the quadratic component of drag force in the yaw direction. r is the heading velocity, $\dot{r}$ is the heading acceleration, N is the moment input to the ROV due to the thrusters (T1, T2, T5, and T6), and ψ is the heading angle. The constant heading velocity refers to $\dot{r}$ heading acceleration equal to 0, and the yaw equation of motion becomes:

$$N_{|r|r}|r|r + N_r r = N \tag{3.26}$$

The test was performed to estimate the drag coefficients $N_{|r|r}$ and N_r in the heading equation. Fixed velocity commands are sent to the four thrusters T1, T2, T5, and T6 and the steady-state heading velocity responses were recorded using the DVL sensor. The moment in yaw direction was computed using the thrusts generated by the thrusters multiplied by the moment arm (i.e. distance) from the center of gravity. Figure 3.22 gives the steady-state angular velocity as a result of the input command to the thrusters (1.3–2.7 V). The plot of the moment versus the angular velocity is shown in Fig. 3.23, where the linear and quadratic damping terms can be determined from the curve fitting. The values of $N_{|r|r}$ and N_r are estimated as 5.4815 and 4.6114, respectively.

The added mass of the ROV in the yaw direction is then identified as shown. A sinusoidal control command (maximum torque corresponding to the heading velocity) is applied to the thrusters T1, T2, T5, and T6 of the ROV. A sinusoidal control signal

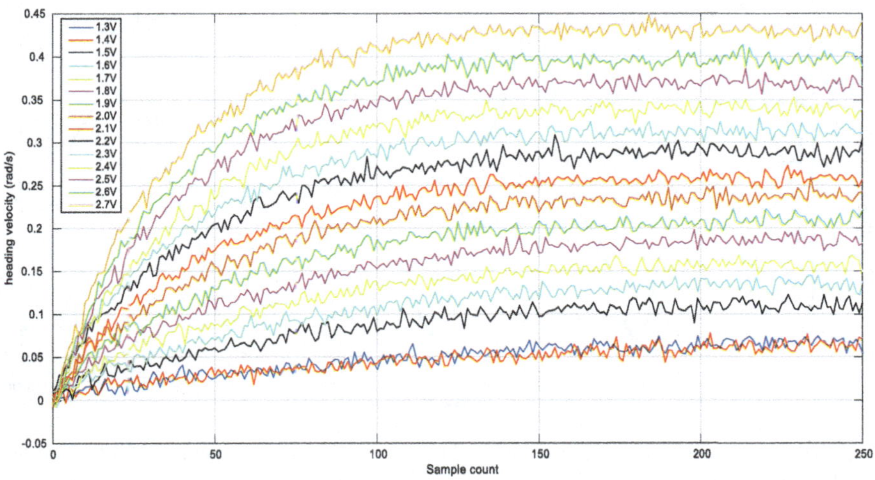

Fig. 3.22 Velocity profile at constant control command to thrusters T1, T2, T5, and T6

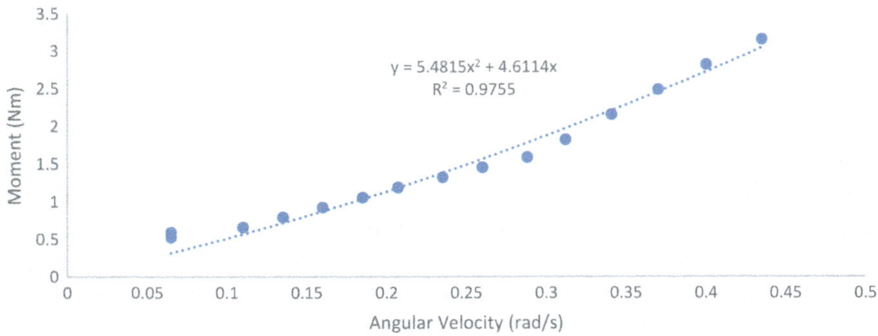

Fig. 3.23 Angular velocity rate versus moment in the yaw direction

with 10 Hz frequency is applied to the thrusters with a value corresponding to +2.7 to −2.7 V (for both forward and reverse direction).

As observed in Figs. 3.24 and 3.25, the maximum heading velocity is observed to be approximately 0.385 rad/s. With 2.7 V control signal into the thruster, the constant heading velocity is approximately 0.435 rad/s that corresponds to 31.52 Nm to achieve this heading speed of 0.435 rad/s. However, the maximum sinusoidal angular speed is around 0.385 rad/s. As the measured heading velocity is smaller than the speed measured in the water tank test, it shows that the ROV needs an additional moment to overcome the resistance due to its body mass and the added mass. If the amplitude of the sine input is increased to around 0.435 rad/s, +2.9 to −2.9 V sine control signal (see Fig. 3.26) has a torque of approximately 35.17 Nm. It indicates

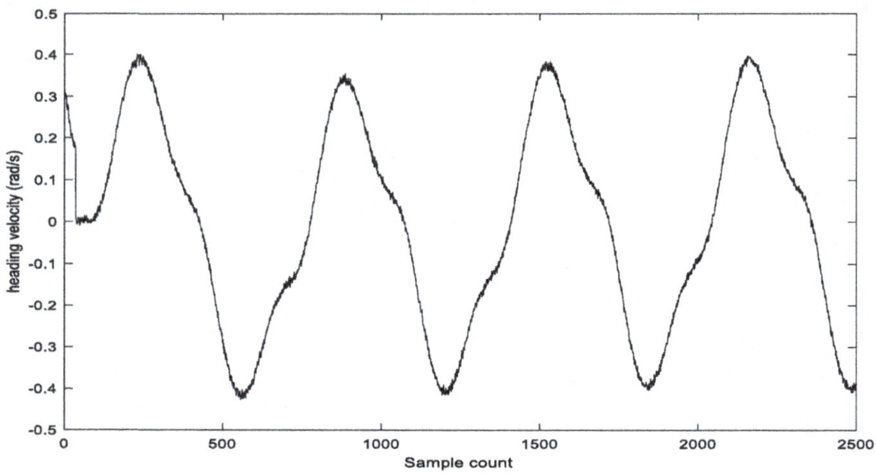

Fig. 3.24 Heading velocity response subjected to sine signal command input (2.7 V) for thrusters T1, T2, T5, and T6

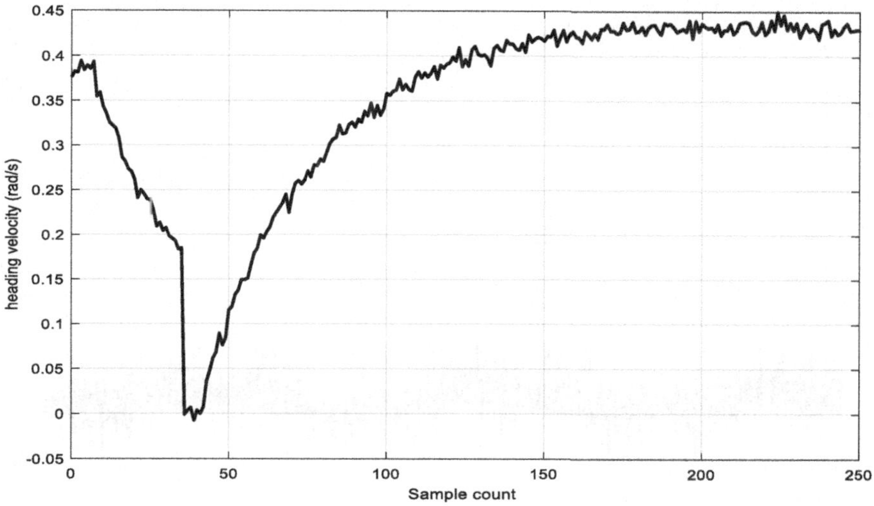

Fig. 3.25 Heading velocity due to constant control command (2.7 V) for thrusters T1, T2, T5, and T6

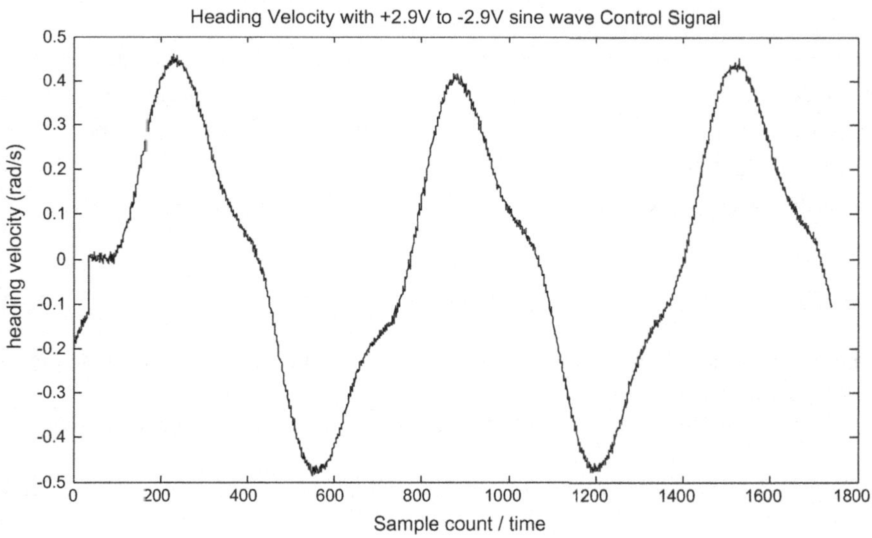

Fig. 3.26 Heading velocity of sine input command of 2.9 V for thrusters T1, T2, T5, and T6

that the ROV needs an additional $35.17 - 31.52 = 3.65$ Nm to overcome both the body inertia and added mass during the motion.

The next step is to compute the angular acceleration using the $+2.9$ to -2.9 V sine control signal. The maximum heading acceleration of approximately 0.85 rad/s^2 was measured (see Fig. 3.27) by the inertial measurement unit sensor.

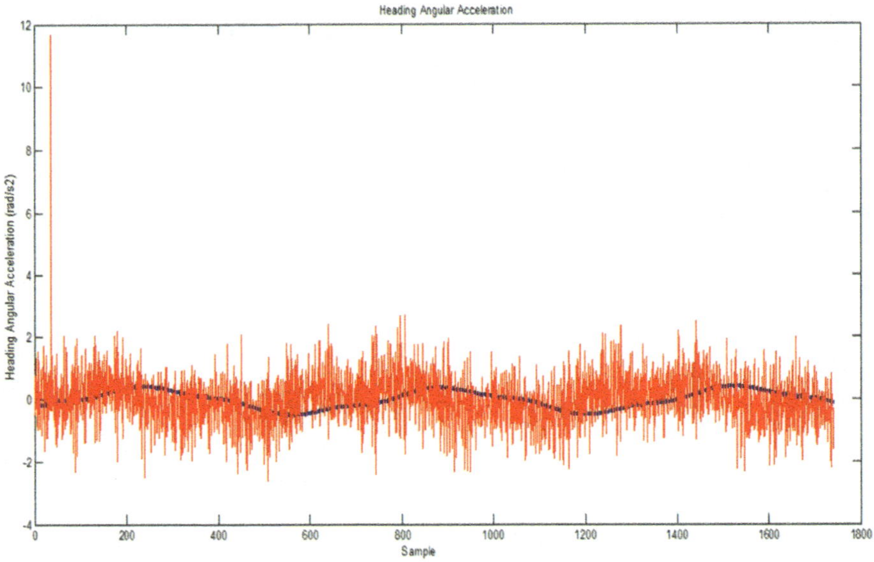

Fig. 3.27 Angular acceleration of sine input signal of 2.9 V

Table 3.10 Comparisons between simulation and experimental results in the yaw direction

Hydrodynamic coefficients (in the yaw direction)	Simulation	Experiment	Error (%)
Added mass (kg)	5.263	4.300	18
Linear damping (N s/rad)	–	4.611	–
Quadratic damping (Nm s^2/rad^2)	6.079	5.481	9.8

$$N_{\dot{r}}\dot{r} = \Delta N \tag{3.27}$$

where ΔN is the difference at the moment and $N_{\dot{r}}$ is the hydrodynamic added mass in the yaw direction.

The moment due to the body inertia and added mass is computed as follows:

$$N_{\dot{r}}\dot{r} = \Delta N = 3.65 \tag{3.28}$$

Substituting the peak acceleration equals 0.85 rad/s^2, $N_{\dot{r}}$ becomes 4.3 kg m^2. As a result, the heading model can be written as:

$$4.3\dot{r} + 5.4815|r|r + 4.6114r = N \tag{3.29}$$

The results tabulated in Table 3.10 show that the quadratic damping has a closer value as compared to the experiments. The added mass does not seem to compare well with the simulation result. It may be due to the experimental test setup in a

Table 3.11 Parameters of ROV

Parameters	Values
Mass	$m = 75$ kg
Weight	$W = 735.75$ N
Buoyancy force	$B = 735.75$ N
Rigid-body mass	$\mathbf{M_{RB}} = \begin{bmatrix} m & 0 & 0 & 0 & mz_G & -my_G \\ 0 & m & 0 & -mz_G & 0 & mx_G \\ 0 & 0 & m & my_G & -mx_G & 0 \\ 0 & -mz_G & my_G & I_x & -I_{xy} & -I_{xz} \\ mz_G & 0 & -mx_G & -I_{yx} & I_y & -I_{yz} \\ -my_G & mx_G & 0 & -I_{zx} & -I_{zy} & I_z \end{bmatrix}$ $= \begin{bmatrix} 75.00 & 0 & 0 & 0 & 0 & 0 \\ 0 & 75.00 & 0 & 0 & 0 & 0 \\ 0 & 0 & 75.00 & 0 & 0 & 0 \\ 0 & 0 & 0 & 2.51 & 0 & 0 \\ 0 & 0 & 0 & 0 & 3.38 & 0.01 \\ 0 & 0 & 0 & 0 & 0.01 & 1.73 \end{bmatrix}$
Added mass	$\mathbf{M_A} = \begin{bmatrix} X_{\dot{u}} & X_{\dot{v}} & X_{\dot{w}} & X_{\dot{p}} & X_{\dot{q}} & X_{\dot{r}} \\ Y_{\dot{u}} & Y_{\dot{v}} & Y_{\dot{w}} & Y_{\dot{p}} & Y_{\dot{q}} & Y_{\dot{r}} \\ Z_{\dot{u}} & Z_{\dot{v}} & Z_{\dot{w}} & Z_{\dot{p}} & Z_{\dot{q}} & Z_{\dot{r}} \\ K_{\dot{u}} & K_{\dot{v}} & K_{\dot{w}} & K_{\dot{p}} & K_{\dot{q}} & K_{\dot{r}} \\ M_{\dot{u}} & M_{\dot{v}} & M_{\dot{w}} & M_{\dot{p}} & M_{\dot{q}} & M_{\dot{r}} \\ N_{\dot{u}} & N_{\dot{v}} & N_{\dot{w}} & N_{\dot{p}} & N_{\dot{q}} & N_{\dot{r}} \end{bmatrix}$ $= - \begin{bmatrix} 20.392 & 0 & 0 & 0 & 0 & 0 \\ 0 & 53.435 & 0 & 0 & 0 & 0 \\ 0 & 0 & 126.144 & 0 & 0 & 0 \\ 0 & 0 & 0 & 2.802 & 0 & 0 \\ 0 & 0 & 0 & 0 & 13.703 & 0 \\ 0 & 0 & 0 & 0 & 0 & 5.263 \end{bmatrix}$
Linear hydrodynamic damping	$\mathbf{D_L} = -\text{diag}\{X_u, Y_v, Z_w, K_p, M_q, N_r\}$ $= -\text{diag}\{[\,3.221\ 3.291\ 5.682\ 0.\ 0\ 0\,]\}$
Quadratic hydrodynamic damping	$\mathbf{D_Q} = -\text{diag}\{X_{u\|u\|}, Y_{v\|v\|}, Z_{w\|w\|}, K_{p\|p\|}, M_{q\|q\|}, N_{r\|r\|}\}$ $= -\text{diag}\ \{[\,105.3\ 139.6\ 273.8\ 0\ 0\ 6.079\,]\}$

(continued)

Table 3.11 (continued)

Parameters	Values
Centre of buoyancy	$x_b = 0,\ y_b = 0,\ z_b = -0.397$
Thrusters' configuration matrix	$\mathbf{T} = \begin{bmatrix} 0 & 0 & 0 & 0 & 0.707 & 0.707 \\ -0.707 & 0.707 & 0 & 0 & 0.707 & -0.707 \\ 0.707 & 0.707 & 1 & 1 & 0 & 0 \\ 0.109 & -0.109 & -0.275 & 0.275 & 0 & 0 \\ 0.278 & 0.278 & -0.035 & -0.035 & 0.304 & 0.304 \\ -0.278 & 0.278 & 0 & 0 & -0.466 & 0.466 \end{bmatrix}$
Centre of gravity	$x_g = 0,\ y_g = 0,\ z_g = 0$

water tank that influences the reading. The effects of the interaction between rotating propellers and propeller-to-ROV's hull have not been included in the simulation.

In summary, the following parameters for the benchmark ROV are tabulated in Table 3.11.

3.7 Simulation of ROV Benchmark Model

The identified ROV parameters from the previous sections are used to model the ROV in MATLAB^{TM}/Simulink^{TM} environment. The differential equation of the ROV is solved by ODE-45. The six inputs (T1–T6) on the left-hand side are the thruster inputs. For example, the ROV is commanded to move in Z-direction using only T3 and T4 thruster. An ROV simulator developed in MATLAB^{TM}/Simulink^{TM} environment is shown in Fig. 3.28. Note that a saturation block diagram is included to limit the voltage inputs to ±100 V. The following steps can be executed to run the ROV model in Fig. 3.28.

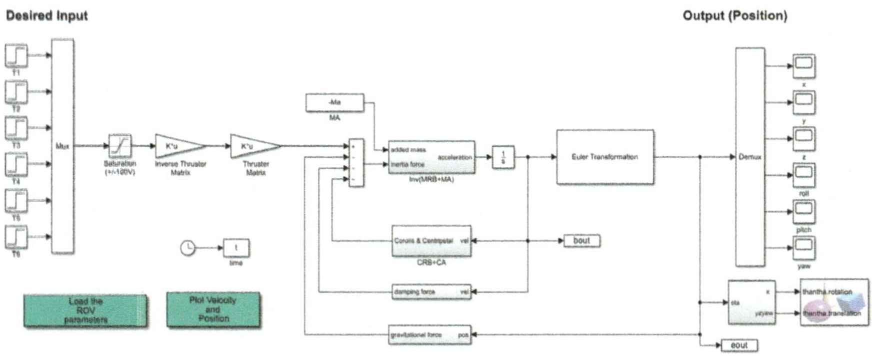

Fig. 3.28 Nonlinear ROV benchmark model

- Load the ROV parameters
- Run the ROV model
- Plot velocity and position after the simulation.

A virtual underwater world environment was developed to enhance the user interface in Fig. 3.26. A virtual reality world for the ROV uses the output signals from the ROV model to move the ROV virtually. It allows the movement and position data to be animated and displayed during the dynamic positioning. First, the CAD model is exported into virtual reality modeling language file format for the V-Realm editor to edit the model. The V-Realm Builder has an extensive object library where the user can import the 3-D background sceneries and objects to create a virtual world. The backgrounds such as the sea and offshore structures are then imported from the library. The VR-sink block diagram in Simulink™ connects between VRML model and Simulink™ block diagrams.

Secondly, the Simulink™ model of the ROV is connected to the virtual world through VR-sink block. By connecting the model to the virtual world, the output data from the Simulink™ model can be used to control and animate the virtual world as shown in Fig. 3.29. The translational ROV's motion outputs and the rotational motion outputs use the Euler's transformation to animate the position of the ROV in the virtual reality world.

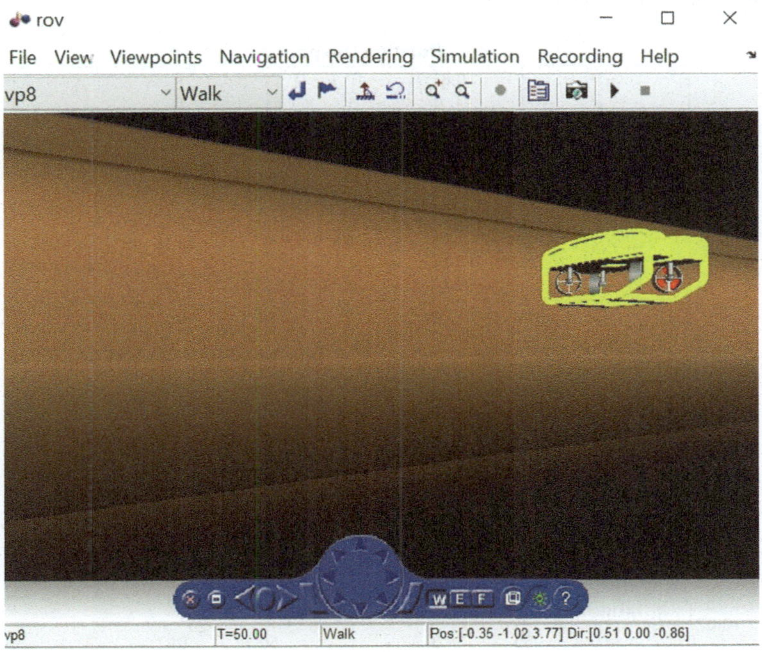

Fig. 3.29 Virtual reality world of ROV

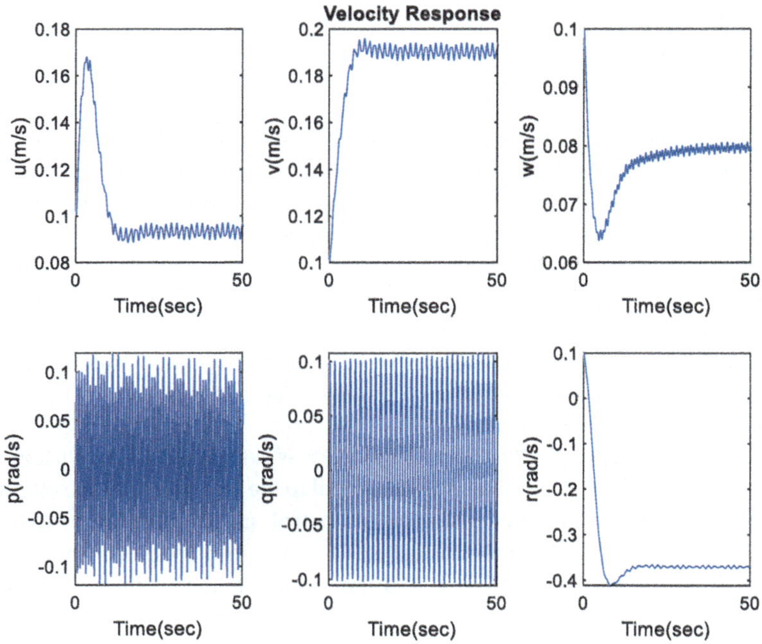

Fig. 3.30 Velocity response of ROV

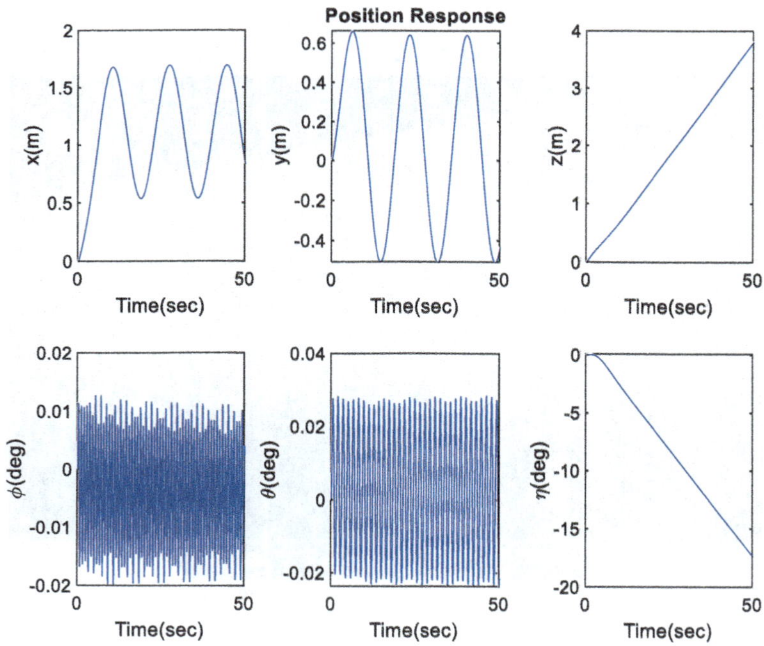

Fig. 3.31 Position response of ROV

The simulated time response of the benchmark ROV model can be seen in Figs. 3.30 and 3.31. The step inputs of 5 V are applied to T1 and T2. The plots show that the ROV velocity is quite oscillatory. The ROV can surge but heave at the same time. The ROV is inherently highly coupled, nonlinear, and uncertain due to the hydrodynamic forces. Thus, it shows that there is a need to have a closed-loop control of the ROV.

3.8 Summary

A systematic modeling of the hydrodynamic damping and added mass of a complex-shaped remotely operated vehicle using a few numerical software was presented. The computational fluid dynamic software STAR-CCM+^{TM} was used to determine the damping parameters of the ROV model. Additionally, potential flow code using WAMIT^{TM} was used to predict the added mass on the ROV model obtained from MULTISURF^{TM} using the panel method to solve the potential flow around the vehicle. The simulated results were verified with the experimental tests in the water tank. Due to test constraints, only the results on the heave and yaw directions were shown. The test results show quite a close match in the added mass for the heave direction and quadratic damping coefficient in the yaw direction. However, the remaining coefficients exhibit some errors as seen in the numerical results. Experimental tests were conducted in a water tank using the joystick as a control to move the ROV to certain desired locations in the heave and yaw directions. The experimental tests exhibit some trends to the simulated results in the heave and yaw directions. In summary, the proposed method provides a viable alternative with reasonable results at an early design stage where the test facilities and workforce can be quite expensive to justify for a prototype ROV. It also provides a sufficient benchmark model and insight into the ROV behavior for better control of the ROV instead of relying on the "black box" approach of using a non-model-based artificial neural network.

The ROV model can be downloaded via the website: https://drive.google.com/open?id=1KkN-LHZ6H-TeIm-K9aaXWIIFtab2jvvA.

References

1. T.I. Fossen, *Marine Control Systems Guidance, Navigation, and Control of Ships* (Rigs and Underwater Vehicles, Marine Cybernetics, Trondheim, Norway, 2002)
2. W.L. Chan, T. Kang, Simultaneous determination of drag coefficient and added mass. IEEE J. Oceanic Eng. **36**(3), 422–430 (2011)
3. O.M. Faltinsen, Hydrodynamics of high-speed marine vehicle (Cambridge University Press, 2006). https://doi.org/10.1017/CBO9780511546068
4. P.R. Wilcox, W.L. Crawford, Least squares method for the reduction of free oscillation data, a NASA Technical Note, NASA TN D-4503 (1968)

5. A.T. Morrison III, D.R. Yoerger, Determination of the hydrodynamic parameters of an underwater vehicle during small scale non-uniform, 1-dimensional translation, in *Proceedings of OCEANS'93* (Victoria, Canada 1993)
6. A. Ross, T.I. Fossen, T.A. Johansen, Identification of underwater vehicle hydrodynamic coefficient using free decay test, in *IFAC Conference on Computer Applications in Marine Systems* (Ancona, Italy, 2004), pp. 363–368
7. Y.H. Eng, W.S. Lau, E. Low, G.G.L. Seet, C.S. Chin, Estimation of the hydrodynamics coefficients of an ROV using free decay pendulum motion, Eng. Lett. **16**(3), EL_16_3_09 (2008)
8. N. Gartner, M. Richier, V. Hugel, Hydrodynamics parameter estimation of submerged bodies: numerical methods comparison and friction model analysis, in *IEEE/RSJ International Conference on Intelligent Robots and Systems (IROS)* (Madrid, Spain, 2018), pp. 5628–5633
9. R. Yang, B. Clement, A. Mansour, H.J. Li, M. Li, N.L. Wu, Modeling of complex-shaped underwater vehicle, in *IEEE International Conference on Autonomous Robot Systems and Competitions (ICARSC)* (Espinho, Portugal, 2014), pp. 36–41
10. C.S. Chin, W.P. Lin, Robust genetic algorithm and fuzzy inference mechanism embedded in sliding-mode controller for uncertain underwater robot, IEEE/ASME Trans. Mechatron. **23**(2), 655–666 (2018)
11. C.S. Chin, W.P. Lin, J.Y. Lin. Experimental validation of open-frame ROV model for virtual reality simulation and control. J. Marine Sci. Technol. **23**(2), 267–287 (2018)

Chapter 4
Control Systems Design of Remotely Operated Vehicle

Abstract Simulation of the theoretical dynamic system is often the first step to obtain an insight into its behaviors before control systems design [1]. If modification of the model is difficult or sometimes infeasible and uncertain, the controller should withstand the model inaccuracy without affecting the performance. The control system design [1] is a combination of experience and techniques. Experience, however, only comes with time as one exposes to more control applications. Usually, the techniques for designing a control system involve mostly a team of engineers.

4.1 Characteristics of ROV for Control

We have published around 25 peer-reviewed publications [2, 4–25] and books [1, 3] as mainly first and corresponding authors (with a controller designed in Simulink^{TM}). The main types of controllers modeled are as follows:

1. Nonlinear PID control with optimization
2. Nonlinear sliding mode control
3. Nonlinear velocity state feedback linearization
4. Nonlinear fuzzy logic control
5. Nonlinear cascaded system control
6. LQG/LTR controller design
7. H-Infinity controller design.

To obtain insight into the ROV's behaviors before control systems design, the following studies on the inherent properties of the benchmark ROV are included.

Before designing a controller for the ROV, inherent properties such as open-loop stability, right-half plane (RHP) zero, and system interactions have to be examined. Since the ROV is unlikely to travel more than $u_o = 2\,\text{m/s}$, $w_o = 2\,\text{m/s}$, $\psi_o =$

Electronic supplementary material The online version of this chapter (https://doi.org/10.1007/978-981-15-6511-3_4) contains supplementary material, which is available to authorized users.

C. S. Chin and M. W. S. Lau, *Benchmark Models of Control System Design for Remotely Operated Vehicles*, https://doi.org/10.1007/978-981-15-6511-3_4

1.57 rad , the following conditions are used. The open-loop ROV system is evaluated at the equilibrium condition: $u_o = 0.001$ m/s, $w_o = 0.05$ m/s, $\psi_o = 1.57$ rad. From Sect. 1.5 (Linear ROV Subsystem Model), the state-space representation of the ROV can be expressed as follows:

$$\dot{\mathbf{x}} = \mathbf{A}(t)\mathbf{x}+\mathbf{B}\tau_{\mathbf{A}} \tag{4.1}$$

$$\begin{bmatrix} \dot{\mathbf{x}}_1 \\ \dot{\mathbf{x}}_2 \end{bmatrix} \overbrace{\begin{bmatrix} -\mathbf{M}^{-1}[\mathbf{C}(t)+\mathbf{D}(t)] & -\mathbf{M}^{-1}\mathbf{G}(t) \\ \mathbf{J}(t) & \mathbf{J}^*(t) \end{bmatrix}}^{\mathbf{A}(t)} \begin{bmatrix} \mathbf{x}_1 \\ \mathbf{x}_2 \end{bmatrix} + \overbrace{\begin{bmatrix} \mathbf{M}^{-1} \\ 0 \end{bmatrix}}^{\mathbf{B}} \tau_{\mathbf{A}} \tag{4.2}$$

where $\mathbf{J}^* = [\mathbf{J}(\eta_0 + \Delta\eta) - \mathbf{J}(\eta_0)]\nu_0$.

```
%MATLAB Script
load('rovmodel75kg.mat')

% Equilibrium point about the vertical plane at a particular yaw angle
uo=0.001;
wo=0.05;
psio=90*pi/180;
Xudot=Ma(1,1);
Zwdot=Ma(3,3);

% Preparing parameters in State-space matrices
ML=M-Ma
DL
CL12=[0 (m(1)-Zwdot)*wo 0;  -(m(1)-Zwdot)*wo 0 (m(1)-Xudot)*uo; 0 -(m(1)-Xudot)*uo 0];
CL=[zeros(3) CL12; -CL12' zeros(3)]
GL=[zeros(3)  zeros(3); zeros(3) [-zb*W 0 0; 0 -zb*W 0; 0 0 0]]
J11=[cos(psio)  -sin(psio) 0; sin(psio) cos(psio) 0; 0 0 1];
J=[J11 zeros(3);zeros(3) eye(3)]

% State-space representation
AA=[-inv(ML)*(CL+DL)]
BB=[inv(ML)*T]
CC=eye(6)
DD=zeros(6,6)
sys_rov=ss(AA,BB,CC,DD);
```

The following state-space matrices can be obtained.

$$\mathbf{A} = \begin{bmatrix} -0.0338 & 0 & 0 & 0 & -0.1054 & 0 \\ 0 & -0.0256 & 0 & 0.0783 & 0 & -0.0007 \\ 0 & 0 & -0.0282 & 0 & 0.0005 & 0 \\ 0 & -1.8933 & 0 & 0 & 0 & 0 \\ 0.5887 & 0 & -0.0056 & 0 & 0 & 0 \\ -0.0008 & 0.0136 & 0 & 0 & 0 & 0 \end{bmatrix};$$

$$\mathbf{B} = \begin{bmatrix} 0 & 0 & 0 & 0 & 0.0074 & 0.0074 \\ -0.0055 & 0.0055 & 0 & 0 & 0.0055 & -0.0055 \\ 0.0035 & 0.0035 & 0.0050 & 0.0050 & 0 & 0 \\ 0.0207 & -0.0207 & -0.0518 & 0.0518 & 0 & 0 \\ 0.0164 & 0.0163 & -0.0021 & -0.0021 & 0.0179 & 0.0178 \\ -0.0399 & 0.0399 & 0 & 0 & -0.0668 & 0.0668 \end{bmatrix};$$

$$\mathbf{C} = \mathbf{I}_6;\ \mathrm{D} = 0$$

The following block diagram (see Fig. 4.1) plots the responses of the linear ROV system (with the following inputs $\bar{u}_3 = \bar{u}_4 = 2$). The inputs will move the ROV within the equilibrium condition.

As compared to the nonlinear response of the ROV, the responses of the linear system are shown in Fig. 4.2. There exists a steady-state error of around 0.6 m/s in heave velocity. The remaining responses match with the linear system.

Open-loop Stability Test: The linear system of ROV is quite stable as seen in Fig. 4.2 where the velocities settle to steady-state values. By examining the eigenvalues of the linear system model, it was found that the system is indeed stable since there are

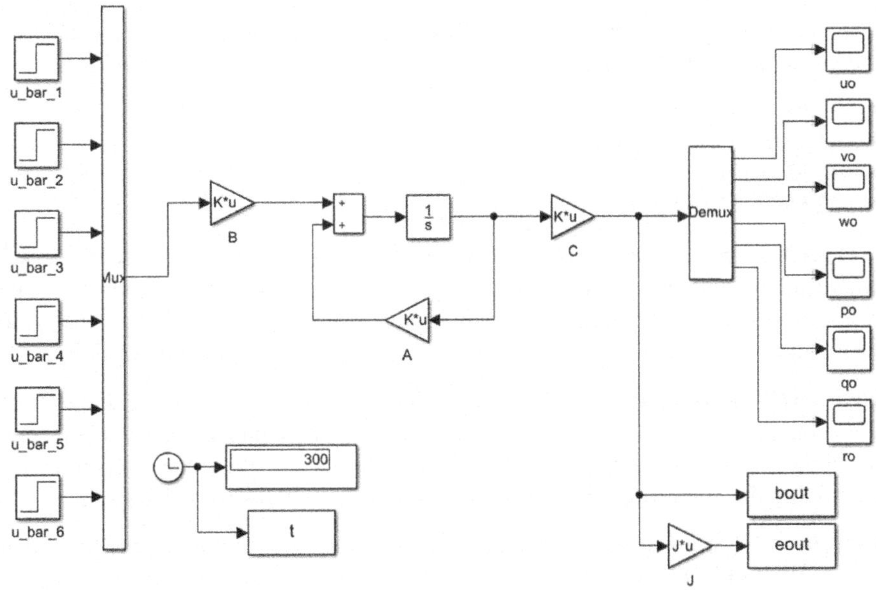

Fig. 4.1 Block diagram of linear open-loop ROV system

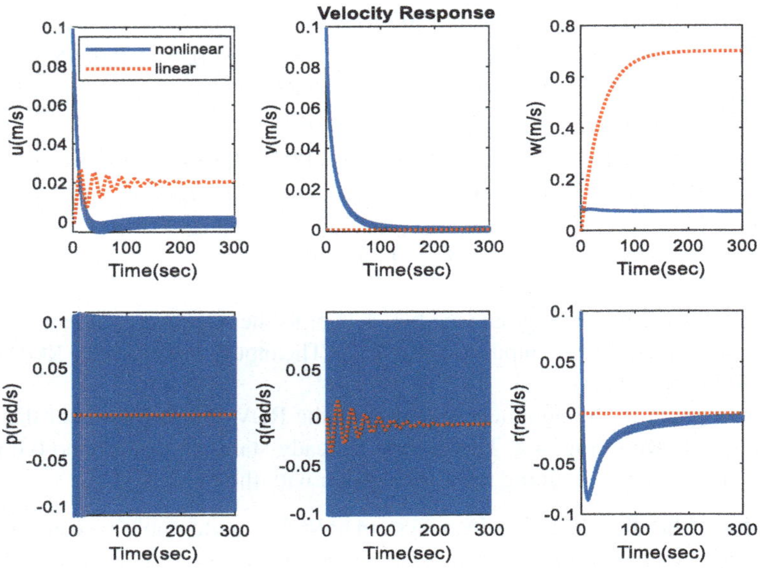

Fig. 4.2 Time response of linear ROV system

no eigenvalues with positive real parts. Besides, the wide disparity in the eigenvalues indicates that the ROV system is quite slow due to the inertia of the ROV.

- $-1.69 \times 10^{-2} \pm 0.246i$
- $-1.28 \times 10^{-2} \pm 0.384i$
- -2.82×10^{-2}
- -2.9835×10^{-19}.

This phenomenon can also be observed in the open-loop time response in Fig. 4.2 where all outputs are settled to their steady-state values as time increased. The responses match the nonlinear system except for the surge velocity. The sway, roll, pitch, and yaw rate are quite small in magnitude and the errors are negligible.

LHP/RHP Zeros Test: Since RHP zeros, in particular, are undesirable due to their effect on the system behavior such as non-minimum phase phenomena and high gain instability in the system time responses, knowledge of their presence is essential. By examining the open-loop RRC ROV system model, no zeros were detected.

Condition Number Test: The condition number indicates that it is quite high, that is, 1×10^{22}. This implies a wide spread of the singular values of the system at certain frequencies. Thus, the ROV system is numerically ill-conditioned.

Controllability/Observability Test: Since the controllability matrix [B AB A^2B ...] has a full rank of 6, the outputs of the ROV are controllable. Similarly, the observability matrix [C; CA; CA^2 ...] has a full rank of 6. It implies that the outputs of the ROV are observable or measurable.

The MATLAB™ commands used are as follows:

```
%MATLAB Script
%eigenstructure
eig_AA_rov=eig(AA)

%transmission zero
trans_zeros_rov=tzero(sys_rov)

%condition number
cond_num_rov=cond(AA)

% Controllability/ observability check
Rank_A_B_rov=rank(ctrb(AA,BB))
Rank_A_C_rov=rank(obsv(AA,CC))
```

Roll and Pitch Stabilizable: Vehicle roll and pitch motions are self-regulating (as seen in Fig. 4.2) as they correspond to the two directions for which it has some degree of self-restoring capabilities provided by the buoyancy and gravitational force vectors forming a restoring couple in roll and pitch directions. As long as the roll and pitch motions are not excessive and the thrusts from all the thrusters reduce to zero, the gravitational and buoyancy moments will eventually restore the ROV to rest state in these two directions. On the other hand, the surges, sway, heave, and yaw motions are not self-regulating and hence left to be controlled in the later section.

4.2 Proportional–Integral–Derivative Control Design for ROV

The proportional–integral–derivative (PID) controller is the most widely used controller structure in industrial applications. Its structural simplicity and sufficient ability to solve many practical control problems have greatly contributed to this wide acceptance. Over the past few decades, many PID design techniques have been proposed for industrial use or even on the ROV. Most of these design techniques are based on simple characterizations of process dynamics, such as the characterization by a first-order model with time delay. Despite this, for plants having higher order such as ROV, there are very few generally accepted design methods that exist. Robust performance design is one of the fundamental problems in control. The problem of robust performance design is to synthesize a controller for which the closed-loop system is internally stabilized and the desired performance specifications are satisfied despite plant uncertainty.

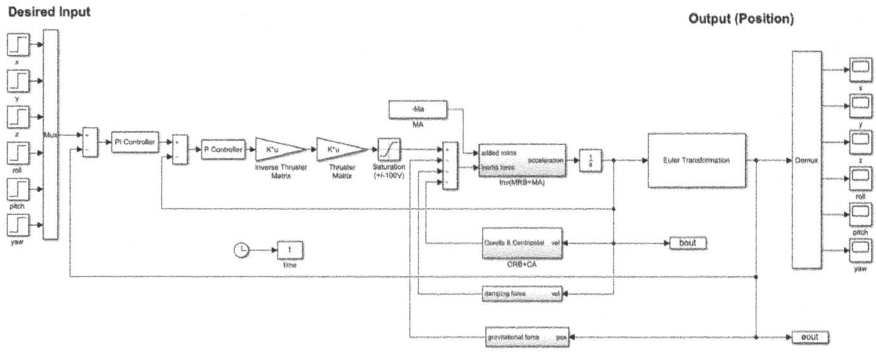

Fig. 4.3 PID–P cascaded control system for ROV

Most existing ROV systems use a series of single-input single-output (SISO) controllers of PID type where each controller is designed for one DOF. In this case, the ROV has six DOFs and hence it should have six PID controllers. This implies that the controller gains matrices $\mathbf{K_p}$, $\mathbf{K_d}$, $\mathbf{K_i}$ in the PID control law becomes:

$$\mathbf{u} = \mathbf{T}^{+}\left(\mathbf{K_p}\mathbf{e}(t) + \mathbf{K_d}\mathbf{e}(t) + \mathbf{K_i}\int_{0}^{t}\mathbf{e}(\tau)d\tau\right) \tag{4.3}$$

where $\mathbf{e}(t)$ is the position error vector and $\mathbf{T}^{+}$ is the Moore–Penrose pseudoinverse matrix. The pseudoinverse can be used for ROV where the thruster configuration matrix $\mathbf{T}$ is non-square as there is equal or more control inputs than controllable DOF. It represents an optimal distribution of control energy for each DOF. The simplified result is computed using $\mathbf{T}^{\mathbf{T}}(\mathbf{T}\mathbf{T}^{\mathbf{T}})^{-1}$.

In Simulink^{TM}, the ROV model with PID and P cascaded controller can be modeled as shown in Fig. 4.3. The variable-step ODE45 solver is used. The PID controller is used to control the position while the P controller is used for the velocity regulation. Here, the P controller can reduce the velocities of the ROV such that the desired positions of the ROV can be achieved. As shown in Fig. 4.3, the PID–P cascaded controller is modeled using Simulink^{TM}. The saturation block diagram is used to limit the thrust outputs from the thrusters. The six inputs on the left-hand side are the desired position inputs. In this simulation, the ROV was commanded to move diagonally in the X–Y plane, that is, $x_d = y_d = 2$ m. The controllers receive the error signals and provide the control action to the ROV. The saturation limit for the thrusters is set as ± 100 V.

The respective outer PID controller gains used for the six degrees of freedom are as follows:

- Surge position: $K_{p,1} = 2$, $K_{d,1} = 0.5$, $K_{i,1} = 6.0$;
- Sway position: $K_{p,2} = 2$, $K_{d,2} = 0.5$, $K_{i,2} = 6.0$;
- Heave position: $K_{p,3} = 3$, $K_{d,3} = 0.5$, $K_{i,3} = 6.0$;
- Roll angle: $K_{p,4} = 3$, $K_{d,4} = 0.5$, $K_{i,4} = 3.0$;
- Pitch angle: $K_{p,5} = 3$, $K_{d,5} = 0.5$, $K_{i,5} = 3.0$;
- Yaw angle: $K_{p,6} = 2$, $K_{d,5} = 0.5$, $K_{i,6} = 6.0$.

The respective inner P controller gains used for the six degrees of freedom are as follows:

- Surge velocity: $K_{p,1} = 5$;
- Sway velocity: $K_{p,2} = 11$;
- Heave velocity: $K_{p,3} = 12$;
- Roll rate: $K_{p,4} = 12$;
- Pitch rate: $K_{p,5} = 12$;
- Yaw rate: $K_{p,6} = 11$.

The position and velocity–time responses of the ROV can be seen in Figs. 4.4 and 4.5, respectively. The ROV can position itself to the desired inputs. The velocities do

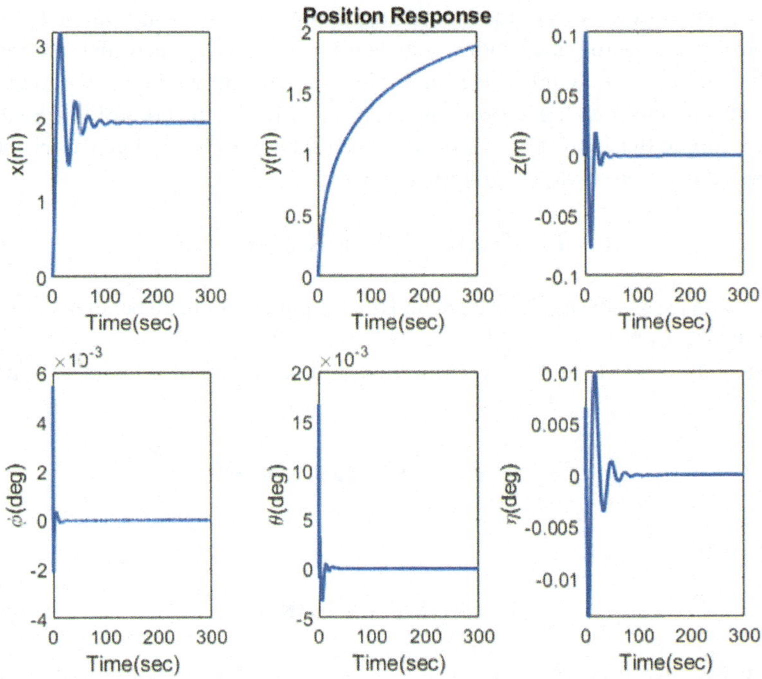

Fig. 4.4 Position responses of ROV using PID controller ($x_d = y_d = 2\,\text{m}$)

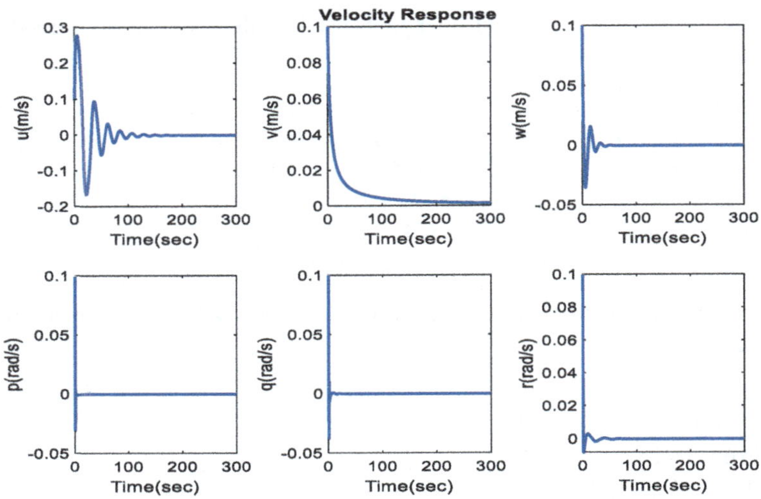

Fig. 4.5 Velocity responses of ROV using PID controller ($x_d = y_d = 2\,\text{m}$)

not exceed 0.2 m/s during the maneuvering. The velocities of the ROV decrease to zero after it reaches the target positions at around 300 s. As seen in the ROV design, the roll and pitch DOFs are quite stable without any control action.

However, most ROV systems for offshore applications use only simple P, PD, and PI controllers for automatic heading and depth control since it is difficult to measure the velocity vector. A standard PID controller can be improved by using the vehicle kinematics together with gravity compensation. Now, let the control law be chosen as just a simple PD control law (instead of the PID controller) where the term $\mathbf{G_f}(\boldsymbol{\eta})$ is included to compensate for gravity and buoyancy, that is:

$$\mathbf{u} = \mathbf{T}^{+}[\mathbf{J}^{\mathrm{T}}(\boldsymbol{\eta})\mathbf{K_p e} - \mathbf{J}^{\mathrm{T}}(\boldsymbol{\eta})\mathbf{K_d J}(\boldsymbol{\eta}) + \mathbf{G_f}(\boldsymbol{\eta})] \tag{4.4}$$

where $\mathbf{e} = \boldsymbol{\eta}_{\mathbf{d}} - \boldsymbol{\eta}$ is the tracking error and the PD gains should be positive or diagonal (if the matrices are used).

This control law is motivated from time differentiation of a Lyapunov function candidate:

$$V(\boldsymbol{\nu}, \mathbf{e}) = \frac{1}{2}(\boldsymbol{\nu}^{\mathrm{T}}\mathbf{M}\boldsymbol{\nu} + \mathbf{e}^{\mathrm{T}}\mathbf{K_p e}) \tag{4.5}$$

which yields

$$\dot{V} = \boldsymbol{\nu}^{\mathrm{T}}[\mathbf{M}\dot{\boldsymbol{\nu}} - \mathbf{J}^{\mathrm{T}}(\boldsymbol{\eta})\mathbf{K_p e}) \tag{4.6}$$

Here we have used the fact that $\dot{\mathbf{e}} = -\dot{\boldsymbol{\eta}} = -\mathbf{J}(\boldsymbol{\eta})\boldsymbol{\nu}$. Substituting the vehicle dynamics into this expression for $\dot{V}$, yields:

$$\dot{V} = \mathbf{v}^{\mathrm{T}}[\mathbf{Tu} - \mathbf{D}(\mathbf{v})\mathbf{v} - \mathbf{G_f}(\mathbf{\eta}) - \mathbf{J}^{\mathrm{T}}(\mathbf{\eta})\mathbf{K_p e}] \tag{4.7}$$

Notice that $\mathbf{v}^{\mathrm{T}}\mathbf{C}(\mathbf{v})\mathbf{v} = 0$ for all $\mathbf{v} \in \Re^6$. From this, it is observed that the proposed PD control law with appropriate choices of $\mathbf{K_p} = \mathbf{K_p^T} > 0$ and $\mathbf{J}^{\mathrm{T}}\mathbf{K_d J} > 0$ ensures that:

$$\dot{V} = -\mathbf{v}^{\mathrm{T}}[\mathbf{D}(\mathbf{v}) + \mathbf{J}^{\mathrm{T}}(\mathbf{\eta})\mathbf{K_d J}(\mathbf{\eta})]\mathbf{v} \leq 0 \tag{4.8}$$

This means that the power is dissipated passively by the damping matrix $\mathbf{D}$ and actively by virtual damping matrix $\mathbf{J}^{\mathrm{T}}(\mathbf{\eta})\mathbf{K_d J}(\mathbf{\eta})$. We now only have to check that the system cannot get stuck at V equal to zero, whenever $\mathbf{e} \neq 0$. From the preceding equation, we see that $\dot{V} = 0$, which implies that $\mathbf{v} = 0$. Hence, substituting the control law into the vehicle dynamics yields:

$$\dot{\mathbf{v}} = \mathbf{M}^{-1}\mathbf{J}^{\mathrm{T}}(\mathbf{\eta})\mathbf{K_p e} \tag{4.9}$$

Consequently, $\dot{\mathbf{v}}$ will be non-zero if $\mathbf{e} \neq 0$ and $\dot{V} = 0$ only if $\mathbf{e} = 0$. Therefore, the system cannot get stuck and the system state vector $\mathbf{\eta}$ will always converge to $\mathbf{\eta_d}$ given $V \to 0$.

As shown in the above equations, Euler's transformation matrix requires transpose and inverse operations. Hence, the transpose of the Euler's transformation matrix is given by:

$$\mathbf{J}^{\mathrm{T}}(\mathbf{\eta_2}) = \begin{bmatrix} \mathbf{J}_1^{\mathrm{T}}(\mathbf{\eta_2}) & 0 \\ 0 & \mathbf{J}_2^{\mathrm{T}}(\mathbf{\eta_2}) \end{bmatrix} \tag{4.10}$$

where $c(.) = \cos$, $s(.) = \sin$, $t(.) = \tan$

$$\mathbf{J}_1^{\mathrm{T}}(\mathbf{\eta_2}) = \begin{bmatrix} \mathrm{c}(\psi)\mathrm{c}(\theta) & \mathrm{s}(\psi)\mathrm{c}(\theta) & -\mathrm{s}(\theta) \\ -\mathrm{s}(\psi)\mathrm{c}(\phi) + \mathrm{c}(\psi)\mathrm{s}(\theta)\mathrm{s}(\phi) & \mathrm{c}(\psi)\mathrm{c}(\phi) + \mathrm{s}(\phi)\mathrm{s}(\theta)\mathrm{s}(\psi) & \mathrm{c}(\theta)\mathrm{s}(\phi) \\ \mathrm{s}(\psi)\mathrm{s}(\phi) + \mathrm{c}(\psi)\mathrm{c}(\phi)\mathrm{s}(\theta) & -\mathrm{c}(\psi)\mathrm{s}(\phi) + \mathrm{s}(\theta)\mathrm{s}(\psi)\mathrm{c}(\theta) & \mathrm{c}(\theta)\mathrm{c}(\phi) \end{bmatrix}$$

$$\mathbf{J}_2^{\mathrm{T}}(\mathbf{\eta_2}) = \begin{bmatrix} 1 & 0 & 0 \\ \mathrm{s}(\phi)\mathrm{t}(\theta) & \mathrm{s}(\phi) & \frac{\mathrm{s}(\phi)}{\mathrm{c}(\theta)} \\ \mathrm{c}(\phi)\mathrm{t}(\theta) & -\mathrm{s}(\phi) & \frac{\mathrm{c}(\phi)}{\mathrm{c}(\theta)} \end{bmatrix} \tag{4.11}$$

The inverse of the Euler's transformation matrix yields:

$$\mathbf{J}^{-1}(\mathbf{\eta_2}) = \begin{bmatrix} \mathbf{J}_1^{-1}(\mathbf{\eta_2}) & 0 \\ 0 & \mathbf{J}_2^{-1}(\mathbf{\eta_2}) \end{bmatrix}; \ \mathbf{J}_1^{-1}(\mathbf{\eta_2}) = \mathbf{J}_1^{\mathrm{T}}(\mathbf{\eta_2}) \tag{4.12}$$

where

$$\mathbf{J}_1^{-1}(\mathbf{\eta}_2)=\begin{bmatrix} c(\psi)c(\theta) & s(\psi)c(\theta) & -s(\theta) \\ -s(\psi)c(\phi)+c(\psi)s(\theta)s(\phi) & c(\psi)c(\phi)+s(\phi)s(\theta)s(\psi) & c(\theta)s(\phi) \\ s(\psi)s(\phi)+c(\psi)c(\phi)s(\theta) & -c(\psi)s(\phi)+s(\theta)s(\psi)c(\theta) & c(\theta)c(\phi) \end{bmatrix}$$

$$\mathbf{J}_2^{-1}(\mathbf{\eta}_2)=\begin{bmatrix} 1 & 0 & -s(\theta) \\ 0 & c(\phi) & c(\theta)s(\phi) \\ 0 & -s(\phi) & c(\theta)c(\phi) \end{bmatrix} \tag{4.13}$$

The derivative of the Euler's transformation matrix yields:

$$\dot{\mathbf{J}}(\mathbf{\eta}_2)=\begin{bmatrix} \dot{\mathbf{J}}_1(\mathbf{\eta}_2) & 0 \\ 0 & \dot{\mathbf{J}}_2(\mathbf{\eta}_2) \end{bmatrix} \tag{4.14}$$

where

$$\dot{\mathbf{J}}_1(\mathbf{\eta}_2)=\begin{bmatrix} J_{11} & J_{12} & J_{13} \\ J_{21} & J_{22} & J_{23} \\ J_{31} & J_{32} & J_{33} \end{bmatrix} \dot{\mathbf{J}}_2(\mathbf{\eta}_2)=\begin{bmatrix} 0 & J_{45} & J_{46} \\ 0 & J_{55} & J_{56} \\ 0 & J_{65} & J_{66} \end{bmatrix} \tag{4.15}$$

and the following parameters are given by:

$$J_{11}=-s(\psi)\dot{\psi}c(\theta)-c(\psi)s(\theta)\dot{\theta};$$

$$J_{12}=-c(\psi)\dot{\psi}c(\phi)+s(\psi)s(\phi)\dot{\phi}-s(\psi)\dot{\psi}s(\theta)s(\phi)+c(\psi)c(\theta)\dot{\theta}s(\phi)+c(\psi)s(\theta)c(\phi)\dot{\phi};$$

$$J_{13}=c(\psi)\dot{\psi}s(\phi)+s(\psi)c(\phi)\dot{\phi}-s(\psi)\dot{\psi}c(\phi)s(\theta)-c(\psi)s(\phi)\dot{\phi}s(\theta)+c(\psi)c(\phi)c(\theta)\dot{\theta};$$

$$J_{21}=c(\psi)\dot{\psi}c(\theta)+s(\psi)s(\theta)\dot{\theta};$$

$$J_{22}=-s(\psi)\dot{\psi}c(\phi)-c(\psi)s(\phi)\dot{\phi}+c(\phi)\dot{\phi}s(\theta)s(\psi)+s(\phi)c(\theta)\dot{\theta}s(\psi)+s(\phi)s(\theta)c(\psi)\dot{\psi};$$

$$J_{23}=s(\psi)\dot{\psi}s(\phi)-c(\psi)c(\phi)\dot{\phi}+c(\theta)\dot{\theta}s(\psi)c(\phi)+s(\theta)c(\psi)\dot{\psi}c(\phi)-s(\theta)s(\psi)s(\phi)\dot{\phi};$$

$$J_{31}=-c(\theta)\dot{\theta};$$

$$J_{32}=-s(\theta)\dot{\theta}s(\phi)+c(\theta)c(\phi)\dot{\phi};$$

$$J_{33}=-s(\theta)\dot{\theta}c(\phi)-c(\theta)s(\phi)\dot{\phi};$$

$$J_{45}=c(\phi)\dot{\phi}t(\theta)+s(\phi)[1+t^2(\theta)]\dot{\theta};\ J_{46}=-s(\phi)\dot{\phi}t(\theta)+c(\phi)[1+t^2(\theta)]\dot{\theta};$$

$$J_{46} = -s(\phi)\dot{\phi}t(\theta) + c(\phi)[1 + t^2(\theta)]\dot{\theta};$$

$$J_{55} = -s(\phi)\dot{\phi};\; J_{56} = -c(\phi)\dot{\phi};$$

$$J_{65} = \frac{c(\phi)\dot{\phi}}{c(\theta)} + \frac{s(\phi)s(\theta)\dot{\theta}}{c^2(\theta)};\; J_{66} = \frac{s(\phi)\dot{\phi}}{c(\theta)} + \frac{c(\phi)s(\phi)\dot{\theta}}{c^2(\theta)};$$

The block diagram of the PD controller that uses the vehicle kinematics and the gravity compensation is shown in Fig. 4.6. The block diagram is quite similar to the one using the PID except for that the PD controller with the vehicle kinematics represented by the transpose of the Euler's transformation matrix is used. The controller

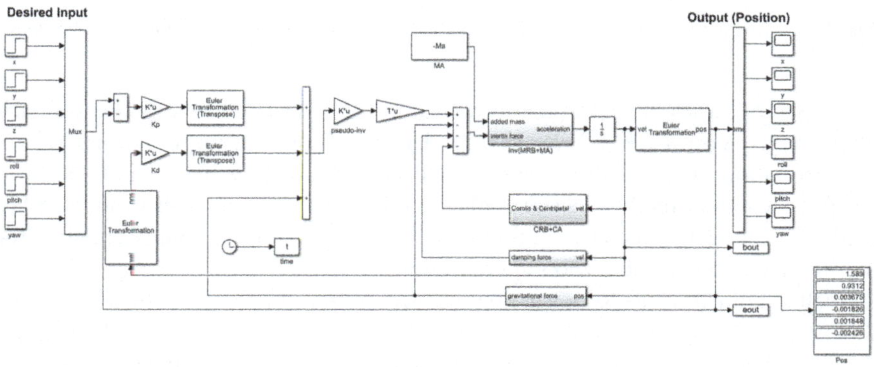

Fig. 4.6 Vehicle kinematics with gravity compensation PID-type control system for ROV

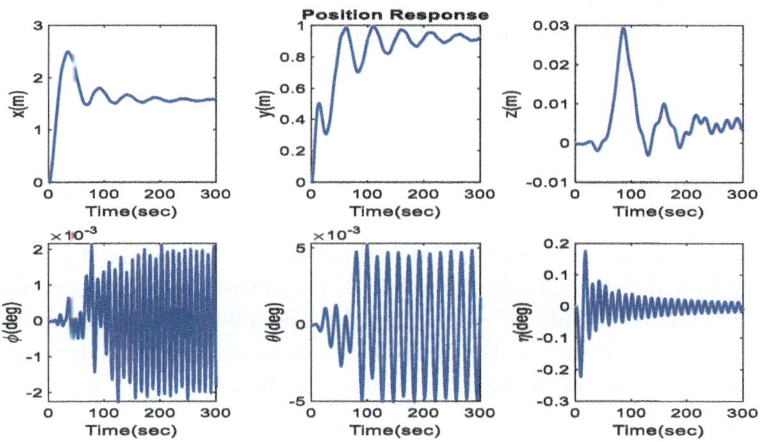

Fig. 4.7 Position time responses of vehicle kinematics with gravity compensation PD-type control ($x_d = y_d = 2$ m)

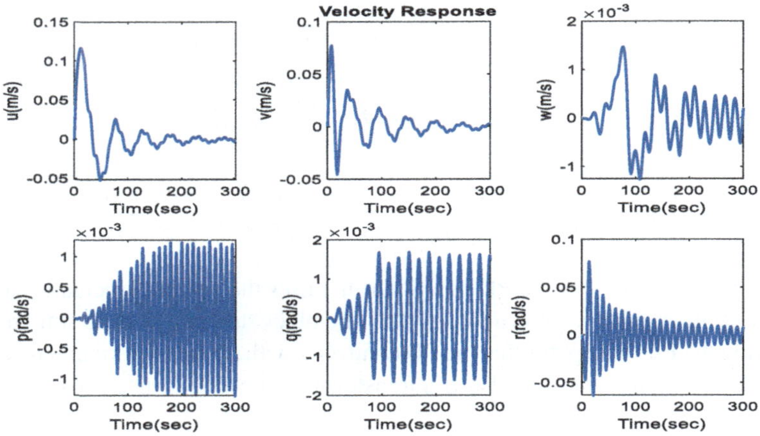

Fig. 4.8 Velocity time responses of vehicle kinematics with gravity compensation PD-type control

block diagram uses the controller as shown in (4.4). The controller gains for each DOF are given as: $K_p = 1.2$, $K_d = 0.75$. The details of the transpose of the Euler's transformation matrix for the translational motion (i.e. $\mathbf{J_1}$ matrix) can be seen in Chap. 1. A similar method can be applied for $\mathbf{J_2}$ matrix as well. This is not the only method for modeling a matrix in Simulink™; however, it is an intuitive method for the simulation and trouble-shooting purposes.

In this simulation, the ROV was commanded to move diagonally in the X–Y plane, that is, $x_d = y_d = 2$ m. By simulating the PD controller that uses the vehicle kinematics and gravity compensation type, the position and velocity–time responses can be seen in Figs. 4.7 and 4.8, respectively. The position and velocity–time responses have improved with less oscillation. But the steady-state error has increased by around 0.5 m and the heave position has a steady-state error of 1 m. The error may be due to the numerical error resulting from the transpose of the Euler's transformation matrix. Further improvement can be made by tuning the controller gains.

4.3 Summary

The control system design using PID-based cascaded controllers on the benchmark ROV model was simulated using Simulink™. The inner-loop controller controls the velocity of the ROV while the outer controller moves the ROV to the desired X–Y positions. The simulations have shown that there exist certain steady-state errors while commanded to move in the horizontal plane direction.

The linear ROV analysis and PID controller for ROV model can be downloaded via the website:

- https://drive.google.com/open?id=1K4KPM7YHClWELnh8DnsPxuSvtB SkzKF1.

4.4 Backstepping Controller Design for ROV

As the roll and pitch motions are asymptotically stable and bounded, they are neglected in the station-keeping model. The station-keeping model involves mainly four DOFs such as: surge, sway, heave, and yaw. In this section, the station-keeping dynamics, followed by the horizontal and vertical plane model that uses only three DOFs are used. The details of formulation of the models can be seen in Sect. 1.5.

Backstepping is a method developed [26, 27] to design stabilizing controls for a certain class of nonlinear systems. Due to the recursive structure, the control designer can begin the design at the known-stable system and have new controllers that progressively stabilize other subsystems. The procedure stops after the final external control is attained.

4.4.1 Backstepping Controller Design for Station-Keeping

Define a virtual state for $\tilde{u}_e$,

$$z_1 = \tilde{u}_e - \alpha_1 \tag{4.16}$$

Choose a control Lyapunov function (CLF) as:

$$V = \frac{1}{2}(m - X_{\dot{u}})\tilde{u}_e^2 \tag{4.17}$$

It gives $\dot{V}$:

$$\dot{V} = -\tilde{u}_e[-X_u\tilde{u}_e + (Y_{\dot{v}} - m)\tilde{v}_e\tilde{r}_e - \tau_x + \bar{\mathrm{u}}_{\mathrm{Ix}}] \tag{4.18}$$

Choose a virtual control as:

$$\alpha_1 = K_u\tilde{u}_e \tag{4.19}$$

Virtual state equation becomes:

$$\dot{z}_1 = -\frac{(1 - K_u)}{m - X_{\dot{u}}}[X_u\tilde{u}_e - (Y_{\dot{v}} - m)\tilde{v}_e\tilde{r}_e + \tau_x - \bar{\mathrm{u}}_{\mathrm{Ix}}] \tag{4.20}$$

Choose an augmented CLF as:

$$V_a = V + \frac{1}{2}z_1^2 \tag{4.21}$$

It gives $\dot{V}_a$:

$$\dot{V}_a = \frac{[\tilde{u}_e(m - X_{\dot{u}}) + (\tilde{u}_e - K_u\tilde{u}_e)(1 - K_u)]}{m - X_{\dot{u}}}[X_u\tilde{u}_e - (Y_{\dot{v}} - m)\tilde{v}_e\tilde{r}_e + \tau_x - \bar{\mathrm{u}}_{\mathrm{Ix}}] \tag{4.22}$$

Actual control $\bar{\mathrm{u}}_{\mathrm{Ix}}$:

$$\bar{\mathrm{u}}_{\mathrm{Ix}} = \frac{(m - X_{\dot{u}})(1 + Y_{\dot{v}}\tilde{v}_e\tilde{r}_e)}{\tilde{u}_e(m - X_{\dot{u}} + 1 - 2K_u + K_u^2)} \tag{4.23}$$

As a result, $\dot{V}_a$ in (4.22) becomes

$$\dot{V}_a = -[X_u + \tilde{u}_e + 1 + \tau_x + \tilde{r}_e\tilde{v}_e m] \tag{4.24}$$

Repeat the same steps in (4.16) to (4.24) for the other virtual state $\tilde{v}_e$, $\tilde{w}_e$, and $\tilde{r}_e$ using $\alpha_2 = K_v\tilde{v}_e$, $\alpha_3 = K_w\tilde{w}_e$, and $\alpha_4 = K_r\tilde{r}_e$, respectively. The actual inner controls for the virtual state $z_2 = \tilde{v}_e - \alpha_2$, $z_3 = \tilde{w}_e - \alpha_3$, and $z_4 = \tilde{r}_e - \alpha_4$ can be expressed as follows:

$$\bar{\mathrm{u}}_{\mathrm{Iy}} = \frac{(m - Y_{\dot{v}})(1 + m\tilde{u}_e\tilde{r}_e)}{\tilde{v}_e(m - Y_{\dot{v}} + 1 - 2K_v + K_v^2)} \tag{4.25}$$

$$\bar{\mathrm{u}}_{\mathrm{Iz}} = \frac{m - Z_{\dot{w}}}{\tilde{w}_e(m - Z_{\dot{w}} + 1 - 2K_w + K_w^2)} \tag{4.26}$$

$$\bar{\mathrm{u}}_{\mathrm{I}\psi} = \frac{(I_z - N_{\dot{r}})(1 + X_{\dot{u}}\tilde{u}_e\tilde{v}_e)}{\tilde{r}_e(I_z - N_{\dot{r}} + 1 - 2K_r + K_r^2)} \tag{4.27}$$

4.4.2 Backstepping Controller Design for Horizontal Plane

Repeat the same procedure in Sect. 4.4.1 for the virtual state $\tilde{u}_e$, $\tilde{v}_e$, and $\tilde{r}_e$ using $\alpha_1 = K_u\tilde{u}_e$, $\alpha_2 = K_v\tilde{v}_e$, and $\alpha_3 = K_r\tilde{r}_e$, respectively. The actual inner controls for the virtual state $z_1 = \tilde{u}_e - \alpha_1$, $z_2 = \tilde{v}_e - \alpha_2$, and $z_3 = \tilde{r}_e - \alpha_3$ can be written as follows:

$$\bar{\mathrm{u}}_{\mathrm{Ix}} = \frac{(m - X_{\dot{u}})(1 + Y_{\dot{v}}\tilde{v}_e\tilde{r}_e)}{\tilde{u}_e(m - X_{\dot{u}} + 1 - 2K_u + K_u^2)} \tag{4.28}$$

$$\bar{u}_{Iy} = \frac{(m - Y_{\dot{v}})(1 + m\tilde{u}_e\tilde{v}_e)}{\tilde{v}_e(m - Y_{\dot{v}} + 1 - 2K_v + K_{\dot{v}}^2)} \tag{4.29}$$

$$\bar{u}_{I\psi} = \frac{(I_z - N_{\dot{r}})(1 + X_{\dot{u}}\tilde{u}_e\tilde{v}_e)}{\tilde{r}_e(I_z - N_{\dot{r}} + 1 - 2K_r + K_{\dot{r}}^2)} \tag{4.30}$$

4.4.3 Backstepping Control System Design for Vertical Plane

Repeat the same procedure in Sect. 4.4.1 for the virtual state $\tilde{u}_e$, $\tilde{w}_e$, and $\tilde{q}_e$ using $\alpha_1 = K_u\tilde{u}_e$, $\alpha_2 = K_w\tilde{w}_e$, and $\alpha_3 = K_q\tilde{q}_e$, respectively. The actual inner controls for the virtual state $z_1 = \tilde{u}_e - \alpha_1$, $z_2 = \tilde{w}_e - \alpha_2$, and $z_3 = \tilde{q}_e - \alpha_3$ become:

$$\bar{u}_{Ix} = \frac{(m - X_{\dot{u}})(1 + \tilde{w}_e\tilde{q}_e m)}{\tilde{u}_e(m - X_{\dot{u}} + 1 - 2K_u + K_{\dot{u}}^2)} \tag{4.31}$$

$$\bar{u}_{Iz} = \frac{(-m + Z_{\dot{w}})(1 + \tilde{u}_e\tilde{q}_e X_{\dot{u}})}{\tilde{v}_e(-m + Z_{\dot{w}} - 1 + 2K_w - K_{\dot{w}}^2)} \tag{4.32}$$

$$\bar{u}_{I\theta} = \frac{(I_y - M_{\dot{q}})(1 + \tilde{u}_e\tilde{w}_e Z_{\dot{w}})}{\tilde{q}_e(I_y - M_{\dot{q}} + 1 - 2K_q + K_{\dot{q}}^2)} \tag{4.33}$$

4.4.4 Simulation of Backstepping Control System

The entire block diagram of the backstepping control can be constructed as shown in Fig. 4.9. The solver options used are fixed as follows: variable-step type, ODE45

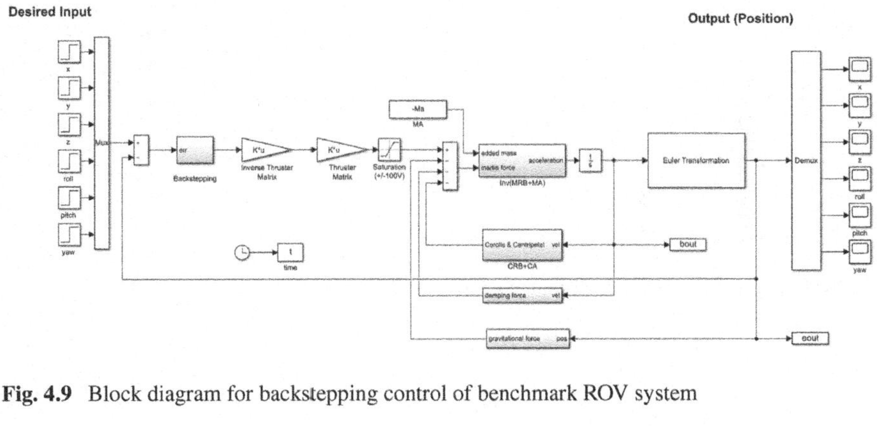

Fig. 4.9 Block diagram for backstepping control of benchmark ROV system

(Dormand-Prince solver and relative tolerance of 0.001). The simulation time is fixed as 300 s. The Simulink™ model includes the backstepping controller block as shown in Fig. 4.10. The backstepping is applied to surge, sway, heave, and yaw directions as seen in Figs. 4.11, 4.12, 4.13 and 4.14, respectively. The roll and pitch angle is not controlled. The equations in (4.23), (4.25) to (4.27) are modeled into the blocks as shown in Figs. 4.11, 4.12, 4.13 and 4.14.

The position and velocity–time responses of the ROV can be seen in Figs. 4.15

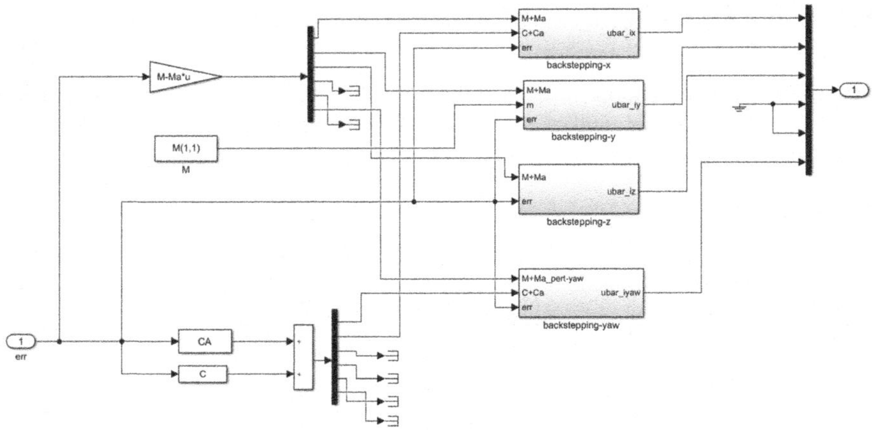

Fig. 4.10 Details of backstepping control block for surge, sway, heave, and yaw directions

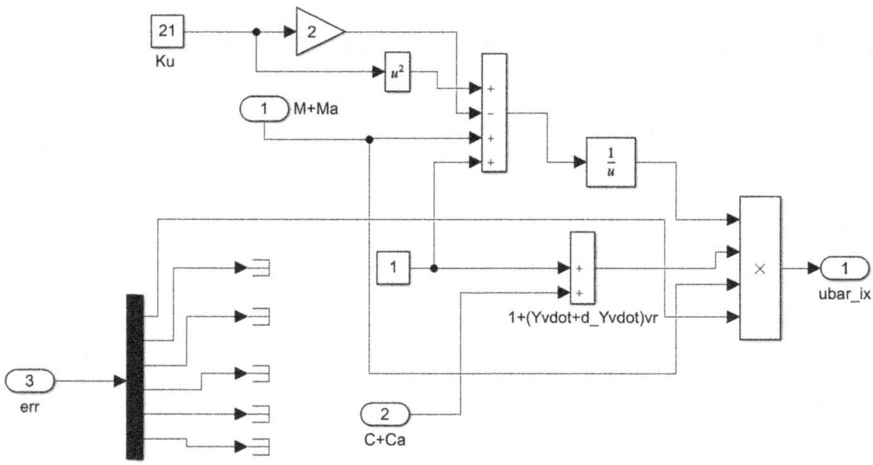

Fig. 4.11 Block diagram for backstepping control in the surge direction

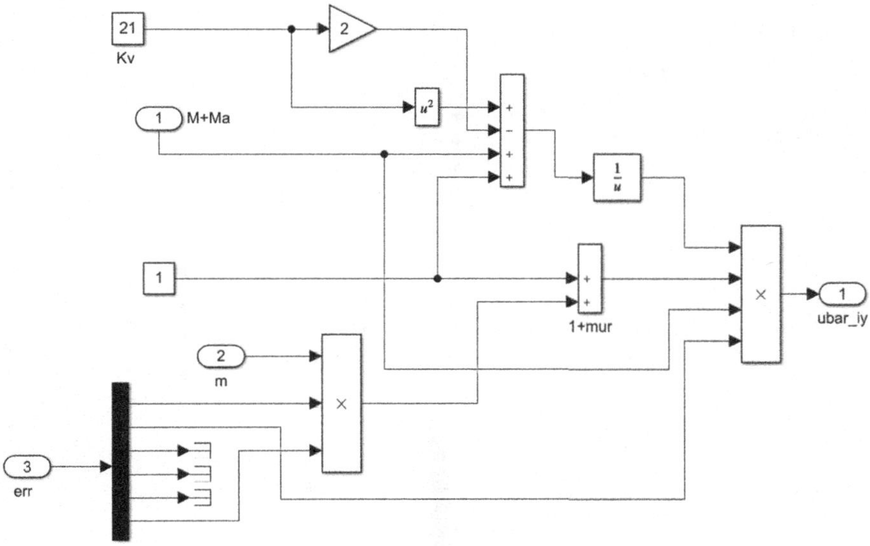

Fig. 4.12 Block diagram for backstepping control in the sway direction

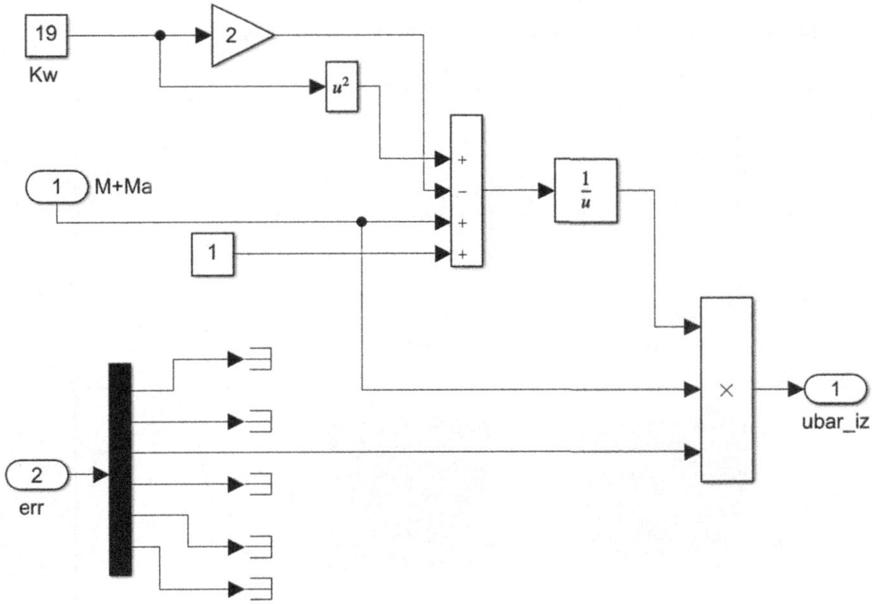

Fig. 4.13 Block diagram for backstepping control in the heave direction

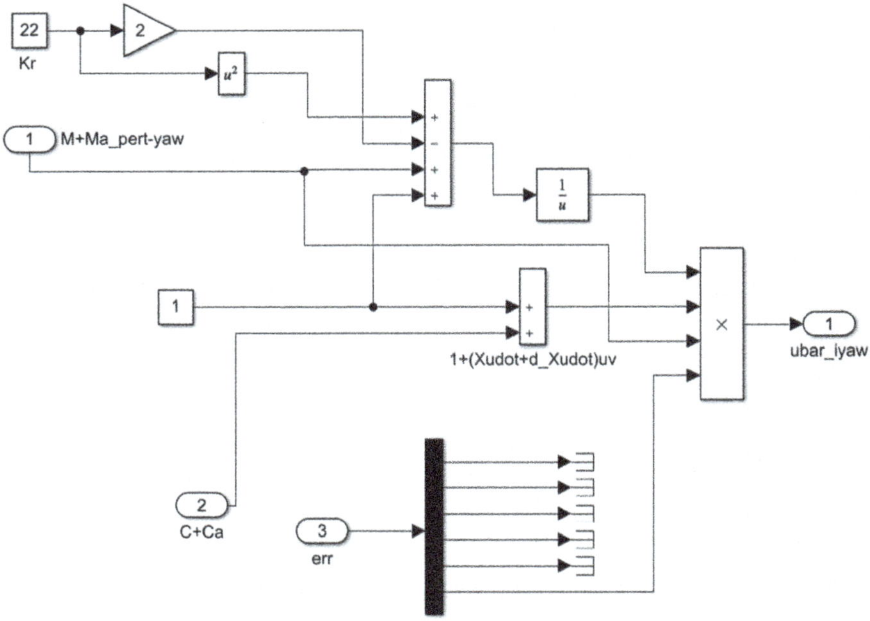

Fig. 4.14 Block diagram for backstepping control in the yaw direction

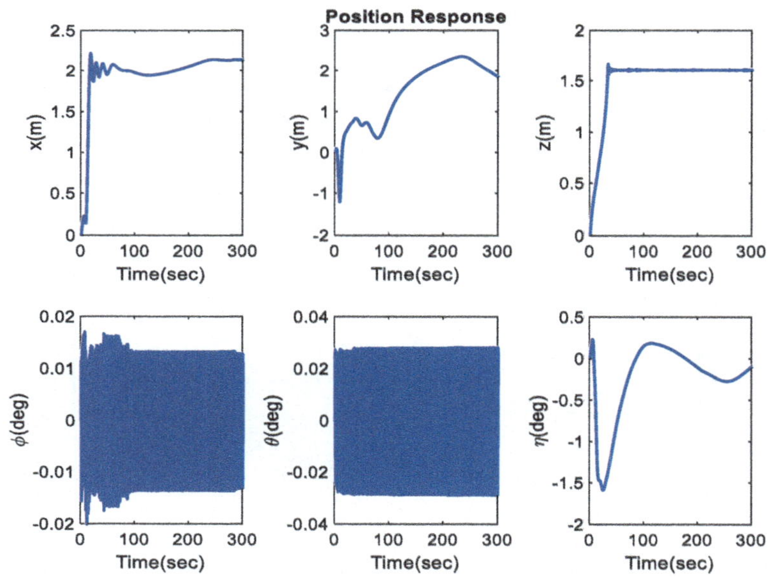

Fig. 4.15 Position response for backstepping control of benchmark ROV system

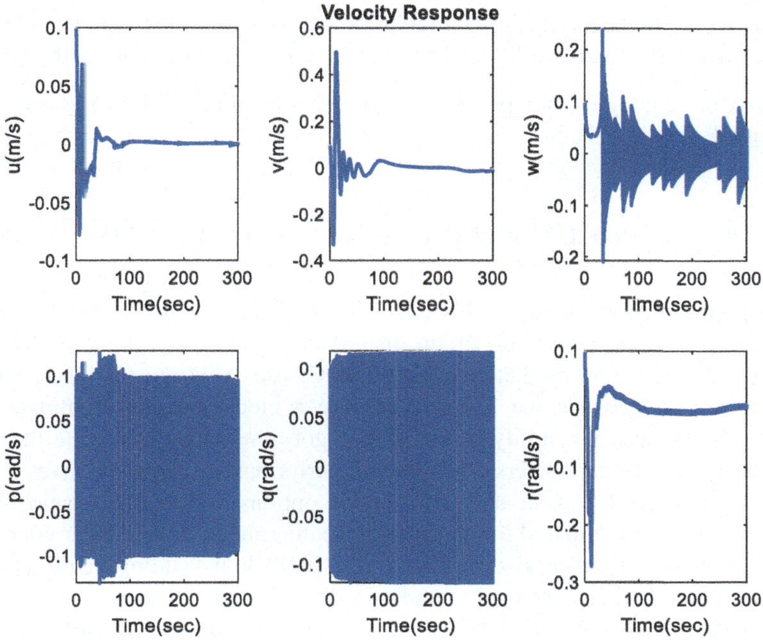

Fig. 4.16 Position responses for backstepping control of benchmark ROV system

and 4.16, respectively. For consistency in comparison, the following desired inputs: $x_d = y_d = 2$ m are used. The ROV exhibits less overshot in surge and sway directions, as compared to the PID controller. The steady-state errors in surge and sway directions are also smaller. However, there exists a larger steady-state error in heave direction. But the velocities of the ROV are quite small such that it will not increase the position. Due to the feedback-stabilizing control law, the position responses are quite stable. The responses also exhibit more damped behavior than the PID controller. As seen in the ROV design, the roll and pitch DOFs are quite self-stabilizing.

4.5 Summary

The backstepping controller was used to design stabilizing controls for the benchmark ROV model. Despite the steady-state error in heave direction, the responses in surge and sway positions give a lower overshoot and steady-state error than the PID controller. The responses also exhibit more damped behavior than the PID controller.

It is mainly due to the feedback-stabilizing control law of the backstepping controller.

The backstepping controller for ROV model can be downloaded via the website:

- https://drive.google.com/open?id=1coQwHlZkcN3nFmXKMajSA2jbaA LJqQ5E.

4.6 Decentralized Model Predictive Control for ROV

Model predictive control on a highly coupled open-frame remotely operated vehicle (ROV) system subjected to uncertain disturbances has always been a challenge. A decentralized model predictive control (MPC) uses a design scaling to balance the interactions between the loops to achieve a block diagonal dominant ROV model. The numerical stability of the MPC algorithm improves despite the sensitivity of the control parameters on the output performance. The MPC gives a better two-error norm performance and more tuning options to control the velocity and position output as compared to other design scaling methods and other controllers such as proportional–integral–derivative (PID), fuzzy-logic, sliding-mode, and PI–P cascaded control when subjected to underwater current disturbance.

Model predictive control (MPC) uses an explicit and separately identifiable model for control. The concept of MPC algorithm is based on the moving horizon approach where the control action is computed to obtain the desired performance over a finite time horizon under uncertainties and constraints [28]. The use of the constraints clearly distinguishes MPC from other control methods, leading to a more reliable controller and tighter control. However, the heavy on-line computational burden [29] and the numerical condition of the control algorithm are the main obstacles in the application of MPC of mostly nonlinear engineering systems. A large computational delay and global minimum or even local minimum cannot be achieved due to time constraints in each time horizon. One approach is to obtain a closed-form optimal MPC [30, 31] to circumvent the computational issue. However, it is hard to predict the system output over a long horizon since the output order is limited by the relative degree of a nonlinear model, and it can be quite unstable [32–34] of a large relative degree. With the derivatives of the control as an additional state, the robustness of the MPC under the model complexity such as highly coupled in states [35] and uncertainty when subjected to external disturbances pose an even greater challenge to the applied control community.

In the current literature, most of the applications of MPC were applied on the following systems such as ALSTOM gasifier control [36], pH control [37], obstacle avoidance for land robot [38], pneumatic brake system [39], trailing edge flapping on wind turbine blade [40], power converters [41], solar energy system [42], induction motor [43], and few autonomous underwater vehicles (AUVs) [44, 45]. The robust stability due to disturbances and the model uncertainties on the AUV were largely studied. However, the model used for the AUV control is quite decoupled in states due to its streamlined body design. The MPC on a highly coupled states model such as an ROV has not been studied. The inherent ROV's model uncertainties due to hydrodynamic parameters, inertial nonlinearity, and coupling between the degrees of freedoms (DoFs) are worsened by the external disturbances due to underwater current. When applying the conventional MPC on this class of underwater vehicles creates a numerical stability issue on the control algorithm.

To overcome the numerical stability problem of MPC on a highly coupled ROV model, we propose the use of non-unity permutation matrices on the ROV model before MPC system design to numerically condition and decentralize the model. It results in a block decentralized [46] model for MPC. This section gives the simulation results of MPC laws and robust stability for the ROV [6]. The application of the MPC with two non-unity permutation matrices on a highly coupled system such as the ROV model before MPC system design leads to a more decentralized predictive control system architecture design and makes the overall approach more decentralized.

4.6.1 Block Diagonal Dominance Model for ROV

Before control system design, another inherent property such as system interaction has to be examined. It helps to understand the couplings between the degrees of freedom of the ROV. The open-loop ROV system is evaluated at the same equilibrium conditions: $u_o = 0.001$ m/s, $w_o = 0.05$ m/s, $\psi_o = 1.57$ rad.

The interaction test using Rosenbrock's row diagonal dominance with Gershgorin discs superimposed on the diagonal elements of the system frequency response (on the transfer function matrix) was performed. It shows the coupling between the states of the ROV.

$$G_{TFM} = \begin{bmatrix} G_{11} & & & & & \\ & G_{22} & & & & \\ & & G_{33} & & & \\ & & & G_{44} & & \\ & & & & G_{55} & \\ & & & & & G_{66} \end{bmatrix} \tag{4.34}$$

The MATLAB^{TM} script forms the transfer function matrix for the linear model of ROV.

MATLAB Script

```
load('rovmodel75kg.mat')

% Equilibrium point about the vertical plane at a yaw angle
uo=0.001;
wo=0.05;
psio=90*pi/180;
Xudot=Ma(1,1);
Zwdot=Ma(3,3);

% Preparing parameters in State-space matrices
ML=M-Ma
DL
CL12=[0 (m(1)-Zwdot)*wo 0;  -(m(1)-Zwdot)*wo 0 (m(1)-Xudot)*uo; 0 -(m(1)-Xudot)*uo 0];
CL=[zeros(3) CL12; -CL12' zeros(3)]
GL=[zeros(3)  zeros(3); zeros(3) [-zb*W 0 0; 0 -zb*W 0; 0 0 0]]
J11=[cos(psio)  -sin(psio) 0; sin(psio) cos(psio) 0; 0 0 1];
J=[J11 zeros(3);zeros(3) eye(3)]

% State-space representation
AA=[-inv(ML)*(CL+DL)]
BB=[inv(ML)*T]
CC=eye(6)
DD=zeros(6,6)

% TF matrix - ROV linear model
[num1 den]=ss2tf(AA,BB,CC,DD,1);[num2 den]=ss2tf(AA,BB,CC,DD,2);
[num3 den]=ss2tf(AA,BB,CC,DD,3);[num4 den]=ss2tf(AA,BB,CC,DD,4);
[num5 den]=ss2tf(AA,BB,CC,DD,5);[num6 den]=ss2tf(AA,BB,CC,DD,6);

G11=tf(num1(1,:),den);G21=tf(num1(2,:),den);G31=tf(num1(3,:),den);G41=tf(num1(4,:),den);
G51=tf(num1(5,:),den);G61=tf(num1(6,:),den);G12=tf(num2(1,:),den);G22=tf(num2(2,:),den);
G32=tf(num2(3,:),den);G42=tf(num2(4,:),den);G52=tf(num2(5,:),den);G62=tf(num2(6,:),den);

G13=tf(num3(1,:),den);G23=tf(num3(2,:),den);G33=tf(num3(3,:),den);G43=tf(num3(4,:),den);
G53=tf(num3(5,:),den);G63=tf(num3(6,:),den);G14=tf(num4(1,:),den);G24=tf(num4(2,:),den);
G34=tf(num4(3,:),den);G44=tf(num4(4,:),den);G54=tf(num4(5,:),den);G64=tf(num4(6,:),den);

G15=tf(num5(1,:),den);G25=tf(num5(2,:),den);G35=tf(num5(3,:),den);G45=tf(num5(4,:),den);
G55=tf(num5(5,:),den);G65=tf(num5(6,:),den);G16=tf(num6(1,:),den);G26=tf(num6(2,:),den);
G36=tf(num6(3,:),den);G46=tf(num6(4,:),den);G56=tf(num6(5,:),den);G66=tf(num6(6,:),den);

% TF matrix
```

We need to ensure that the linear ROV model is at its minimal realization (or at least possible order). The controllability and observability check shows that the linear ROV model is still controllable and observable (i.e. rank is 6). The condition number reduces from 1×10^{22} to 8.5×10^{20}.

MATLAB SCRIPT

```
% reduction of the system (minimal realization)
[p,q,r]=dmperm(AA);
AAred=AA(p,q);
[p,q,r]=dmperm(BB);
BBred=BB(p,q);

% Controllability/ observability & condition number check again
Rank_A_B_rov=rank(ctrb(AAred,BBred))
Rank_A_C_rov=rank(obsv(AAred,CC))
cond(AAred)
```

The Gershgorin discs superimposed on the diagonal elements of the system frequency response on G_{11} are then performed. For example, the script to draw the first element in the transfer function matrix, G_{11} and G_{22} can be seen below. The remaining plots can be performed in the same way.

MATLAB Script

```
%----------G11-----------------------------------------------------------------
figure()
clear i n
box on
drawnow; box on;
subplot(6,6,1);
sw=logspace(-2,10,100);
hold on;
% use one norm or Edmunds scaling

  G_ger_g11=G_tot(1,1);
  [re_g11_1 im_g11_1]=nyquist(G_ger_g11,sw);

 % nyquist
 for xx=1:length(sw)
    re_g11(xx)=re_g11_1(:,:,xx);
    im_g11(xx)=im_g11_1(:,:,xx);
  end
  plot(re_g11,im_g11)

 %ger. disk
  for n=1:length(sw)

  G_ger=evalfr(G_tot,sw(n)*i);
  G11_ger=evalfr(G_tot(1,1),sw(n)*i);
  G22_ger=evalfr(G_tot(2,2),sw(n)*i);
  G33_ger=evalfr(G_tot(3,3),sw(n)*i);
  G44_ger=evalfr(G_tot(4,4),sw(n)*i);
  G55_ger=evalfr(G_tot(5,5),sw(n)*i);
  G66_ger=evalfr(G_tot(6,6),sw(n)*i);
```

```
diag_block_ger=[G11_ger 0   0        0   0   0;
                                     0  G22_ger 0       0   0   0;
        0    0  G33_ger   0   0   0;
        0    0   0   G44_ger   0   0;
        0    0   0        0  G55_ger  0;
        0    0   0        0   0  G66_ger];
  off_diag_block_ger=G_ger-diag_block_ger;

  radius_g11_1(n)=sum(abs(off_diag_block_ger(1,:)));
  radius_g22_2(n)=sum(abs(off_diag_block_ger(2,:)));
  radius_g33_3(n)=sum(abs(off_diag_block_ger(3,:)));
  radius_g44_4(n)=sum(abs(off_diag_block_ger(4,:)));
  radius_g55_5(n)=sum(abs(off_diag_block_ger(5,:)));
  radius_g66_6(n)=sum(abs(off_diag_block_ger(6,:)));

  [lat_g11_1,lon_g11_1]=scircle1(re_g11_1(:,:,n),im_g11_1(:,:,n),radius_g11_1(n));
   plot(lat_g11_1,lon_g11_1);

end
```

```
%---------------G22-----------------------------------------------------------------------------------
subplot(6,6,8);box on;

sw=logspace(-2,10,100);
hold on;

  G_ger_g22=G_tot(2,2);
  [re_g22_1,im_g22_1]=nyquist(G_ger_g22,sw);

 % nyquist
 for xx=1:length(sw)
    re_g22(xx)=re_g22_1(:,:,xx);
    im_g22(xx)=im_g22_1(:,:,xx);
  end
   plot(re_g22,im_g22)

  %ger. disk
   for n=1:length(sw)
  [lat_g33_2,lon_g33_2]=scircle1(re_g22_1(:,:,n),im_g22_1(:,:,n),radius_g22_2(n));
  plot(lat_g33_2,lon_g33_2);

end
```

The Gershgorin discs superimposed on the diagonal elements of the system frequency response plots are shown in Fig. 4.17, which indicates that the system is highly interactive over all frequencies (1–100 rad/s). This can be observed from the increasing size of the discs in the diagonal elements of each row. The growing size of the disc implies that the off-diagonal terms are more dominant than its diagonal terms. To minimize the control effort required for a highly coupled system, pre-conditioning (i.e. to obtain a more diagonal dominance) on the ROV system is favorable to facilitate controller design. Note that to achieve diagonal dominance, either row or column dominance is sufficient.

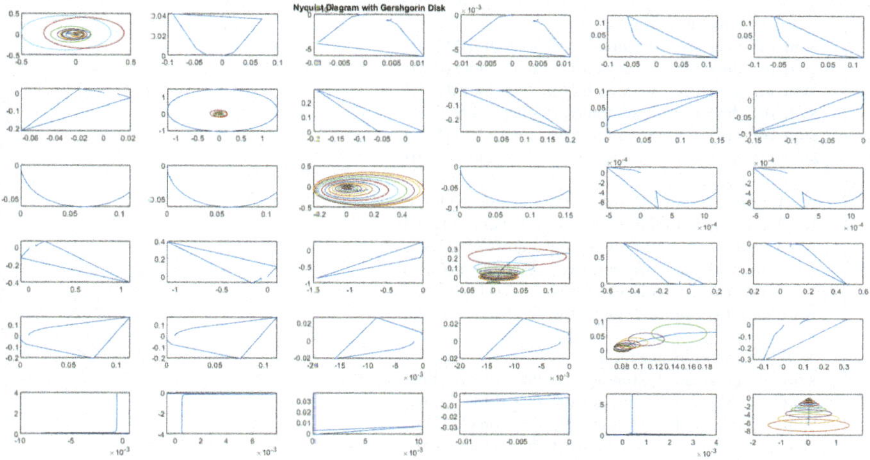

Fig. 4.17 Direct Nyquist array (DNA) with Gershgorin discs before scaling

The corresponding row and column dominance measure (i.e. the sum of the off-diagonal terms in each row or column over the absolute value of the transfer function evaluated at each frequency) shows that the magnitudes are indeed not small. The two-norm values are approximately 3.21×10^3 and 7.58×10^5 for the row and column dominance, respectively. It further confirmed that the ROV is highly coupled in motions.

MATLAB Script

```
% col & row diagonal dominance
rro=100;
table_domin2=cell(rro,7);
table_domin=cell(rro,7);
pp1=1;
s=logspace(-5,2,100);

for n=1:rro
  % min svd of diagonal block / max svd of off-diagonal block--> Block diagonal dominance measure
  G_sv=evalfr(G_tot,s(n)*i);
  diag_block=
          [G_sv(1,1)  0     0     0        0     0;
           0  G_sv(2,2)  0     0        0     0;
           0     0  G_sv(3,3)  0        0     0;
           0     0     0  G_sv(4,4)     0     0;
           0     0     0     0     G_sv(5,5)  0;
           0     0     0     0        0  G_sv(6,6)];

  off_diag_block=G_sv-diag_block;
```

```
% rows dominance ratios
sum_off_diag_row_1=sum(abs(off_diag_block(1,:)));
sum_off_diag_row_2=sum(abs(off_diag_block(2,:)));
sum_off_diag_row_3=sum(abs(off_diag_block(3,:)));
sum_off_diag_row_4=sum(abs(off_diag_block(4,:)));
sum_off_diag_row_5=sum(abs(off_diag_block(5,:)));
sum_off_diag_row_6=sum(abs(off_diag_block(6,:)));

row1_ratio(n)=sum_off_diag_row_1/abs(G_sv(1,1));
row2_ratio(n)=sum_off_diag_row_2/(abs(G_sv(2,2)));
row3_ratio(n)=sum_off_diag_row_3/abs(G_sv(3,3));
row4_ratio(n)=sum_off_diag_row_4/(abs(G_sv(4,4)));
row5_ratio(n)=sum_off_diag_row_5/abs(G_sv(5,5));
row6_ratio(n)=sum_off_diag_row_6/(abs(G_sv(6,6)));

% columns dominance ratios
sum_off_diag_col_1=sum(abs(off_diag_block(:,1)));
sum_off_diag_col_2=sum(abs(off_diag_block(:,2)));
sum_off_diag_col_3=sum(abs(off_diag_block(:,3)));
sum_off_diag_col_4=sum(abs(off_diag_block(:,4)));
sum_off_diag_col_5=sum(abs(off_diag_block(:,5)));
sum_off_diag_col_6=sum(abs(off_diag_block(:,6)));

col1_ratio(n)=sum_off_diag_col_1/abs(G_sv(1,1));
col2_ratio(n)=sum_off_diag_col_2/(abs(G_sv(2,2)));
col3_ratio(n)=sum_off_diag_col_3/abs(G_sv(3,3));
col4_ratio(n)=sum_off_diag_col_4/(abs(G_sv(4,4)));
col5_ratio(n)=sum_off_diag_col_5/abs(G_sv(5,5));
col6_ratio(n)=sum_off_diag_col_6/(abs(G_sv(6,6)));

% create cell to store ratio
table_domin{pp1,1}=s(n);
table_domin{pp1,2}=row1_ratio(n);
table_domin{pp1,3}=row2_ratio(n);
table_domin{pp1,4}=row3_ratio(n);
table_domin{pp1,5}=row4_ratio(n);
table_domin{pp1,6}=row5_ratio(n);
table_domin{pp1,7}=row6_ratio(n);

table_domin2{pp1,1}=s(n);
table_domin2{pp1,2}=col1_ratio(n);
table_domin2{pp1,3}=col2_ratio(n);
table_domin2{pp1,4}=col3_ratio(n);
table_domin2{pp1,5}=col4_ratio(n);
table_domin2{pp1,6}=col5_ratio(n);
table_domin2{pp1,7}=col6_ratio(n);

pp1=pp1+1;

end
```

Also, the singular value and condition number at different frequencies plot show that the ROV is highly coupled in motions. As shown in Fig. 4.18, there are many peaks, in particular at 0.38 rad/s. A design scaling is therefore required to reduce the coupling and to improve the numerical condition using a more decentralized structure before performing MPC.

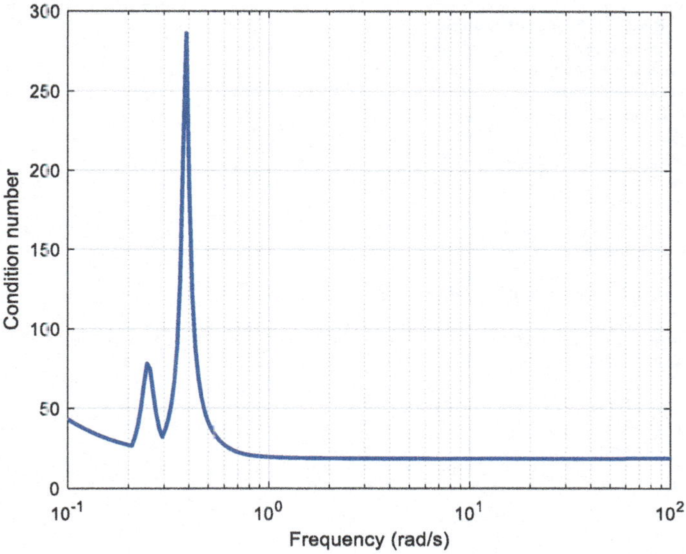

Fig. 4.18 Condition number at different frequency

MATLAB Script

```
figure()
s_tot=pck(AA,BB,CC,DD);
w=logspace(-1,2,200);
sv_tot=sigma(s_tot,w);
semilogx(w,max(sv_tot)./min(sv_tot));
ylabel('Condition number');
xlabel('Frequency (rad/s)');
```

Various design scaling methods such as Edmunds scaling [32], one-norm, and Perron-Frobenius (PF) scaling are used. The scaling frequency at 0.38 rad/s is chosen. The outputs are a non-unity diagonal pre- and post-compensator with different values. The Edmunds scaling algorithm (see Appendix for the algorithm written in MATLAB) provides both scaling and input/output pairings as seen in the non-unity permutation matrices.

MATLAB Script (Edmunds Scaling)

```
function [pre,post,gs]=scale(gin,meth,niter)
error(nargchk(1,3,nargin));
if(nargin<2) meth=2;end
if(nargin<3) niter=30;end
g=abs(gin);
[r,c]=size(g);
dpre=ones(1,c);
makerows=ones(1,r);
makecols=ones(1,c);
gs=gin.*dpre(makerows,:);
dpost=ones(r,1);
gs=dpost(:,makecols).*gin.*dpre(makerows,:);

 % Make each row sum approximately 1 or zero  and each column sum approximately 1 or zero
addrows=ones(c,1);
addcols=ones(1,r);
mindim=min([r c]);
rscale=r/mindim;
cscale=c/mindim;

 % Step 1
gs=dpost(:,makecols).*g.*dpre(makerows,:);
iter=0;
rowsums=(sum(gs'))';
L=cscale*(addcols')./rowsums;
dpost=dpost.*L;

% Step 2
for iter=1:niter
  gs=dpost(:,makecols).*g.*dpre(makerows,:);

  columnsums=sum(gs);

  K=rscale*(addrows')./columnsums;
  dpre=rscale*dpre.*K;
  gs=dpost(:,makecols).*g.*dpre(makerows,:);

  rowsums=(sum(gs'))';
  L=cscale*(addcols')./rowsums;
  dpost=dpost.*L;
end

%gs=post*gin*pre;
logpre=sum(real(log(dpre)))/length(dpre);
logpost=sum(real(log(dpost)))/length(dpost);
adjust=exp((logpre-logpost)/2);
dpre=dpre/adjust;
dpost=dpost*adjust;
gs=dpost(:,makecols).*gin.*dpre(makerows,:);
post=diag(dpost);
pre=diag(dpre);
```

MATLAB Script (One-Norm Scaling)

```
function [pre,post,gs]=scale(gin,meth,niter)

error(nargchk(1,3,nargin));
if(nargin<2) meth=2;end
if(nargin<3) niter=30;end
g=abs(gin);
[r,c]=size(g);
dpre=ones(1,c);
makerows=ones(1,r);
makecols=ones(1,c);
gs=gin.*dpre(makerows,:);

dpost=ones(r,1);
gs=dpost(:,makecols).*gin.*dpre(makerows,:);


%make each row sum approximately 1 or zero  and each column sum approximately 1 or zero
addrows=ones(c,1);
addcols=ones(1,r);
mindim=min([r c]);
rscale=r/mindim;
cscale=c/mindim;

% Step 1
gs=dpost(:,makecols).*g.*dpre(makerows,:);
iter=0;
rowsums=(sum(gs'))';
L=cscale*(addcols')./rowsums;
dpost=dpost.*L;

% Step 2
for iter=1:niter
  gs=dpost(:,makecols).*g.*dpre(makerows,:);
  columnsums=sum(gs);
  K=rscale*(addrows')./columnsums;
  dpre=rscale*dpre.*K;
  gs=dpost(:,makecols).*g.*dpre(makerows,:);
  rowsums=(sum(gs'))';
  L=cscale*(addcols')./rowsums;
  dpost=dpost.*L;
end
%gs=post*gin*pre;
logpre=sum(real(log(dpre)))/length(dpre);
logpost=sum(real(log(dpost)))/length(dpost);
adjust=exp((logpre-logpost)/2);

dpre=dpre/adjust;
dpost=dpost*adjust;
gs=dpost(:,makecols).*gin.*dpre(makerows,:);
post=diag(dpost);
pre=diag(dpre);
```

MATLAB Script (Perron-Frobenius scaling)

```
function [pfval,pre,post,gs]=speron(gin)

error(nargchk(1,1,nargin));
g=abs(gin);
[r,c]=size(g);
gdiag=diag(g);
ze=find(gdiag==0);
% perturb any zero elements on the diagonal
if(length(ze)>0)
  pfval=1.e300;
  pre=zeros(c);
  post=zeros(r);
  gs=zeros(r,c);
  for col=ze
    disp(['Diagonal element ' int2str(col) ' was zero and so the Peron Frobenius Eigenvalue is'])
    disp('infinite, and the scaling does not exist')
  end
 return
end
t=g*inv(diag(gdiag));
[u,l]=eig(t);
[pfval,lnum]=max(diag(l));
if(nargout<2), return,end

pre=inv(diag(gdiag))*abs(diag(u(:,lnum)));
if(nargout<3), return,end
g1=g*pre;
%    to test the dominance    dom=(g1*ones(c,1))./diag(g1)
g1=g1';
gdiag=diag(g1);
t=g1*inv(diag(gdiag));
[u,l]=eig(t);
[pfval,lnum]=max(diag(l));
post=inv(diag(gdiag))*abs(diag(u(:,lnum)));
if(nargout<4), return,end
gs=post*g*pre;
%    to test the dominance    dom=(gs*ones(c,1))./diag(gs)
%    to test the dominance    dom=(ones(1,c)*gs')./diag(gs)'
```

The row dominance ratio of each scaling method is compared. It was found that the Edmunds scaling (row dominance ratio: 900, column dominance ratio: 9,000) has smaller dominance ratio (i.e. smaller off-diagonal terms than the diagonal terms) than the one-norm scaling (row dominance ratio: 850, column dominance ratio: 35,000) and the Perron–Frobenius scaling (row dominance ratio: 550, column dominance ratio: 11,000). The plots for the different scaling methods can be seen in Figs. 4.19, 4.20, 4.21, 4.22, 4.23 and 4.24.

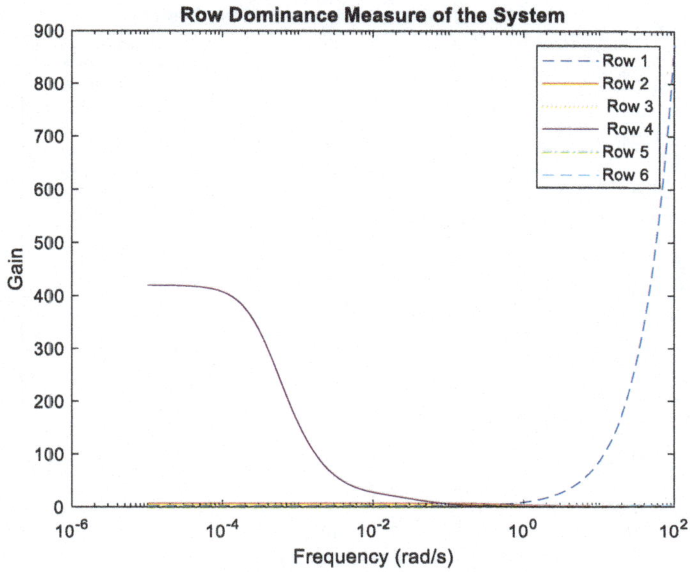

Fig. 4.19 Row dominance for one-norm scaling at different frequency

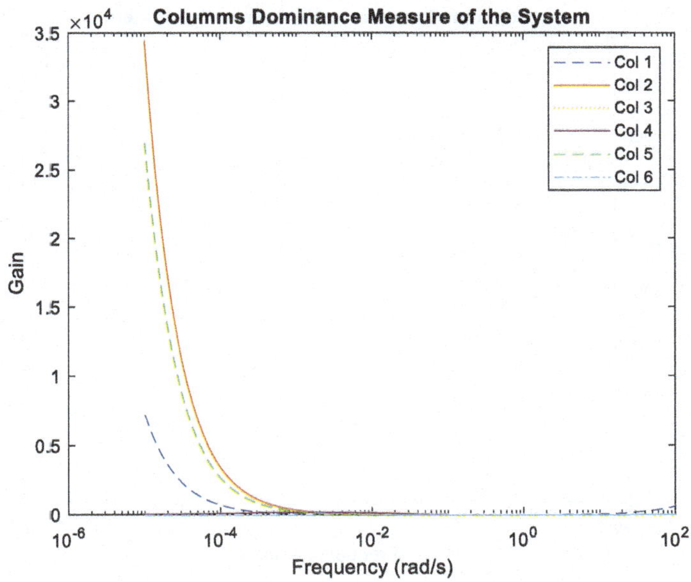

Fig. 4.20 Column dominance for one-norm scaling at different frequency

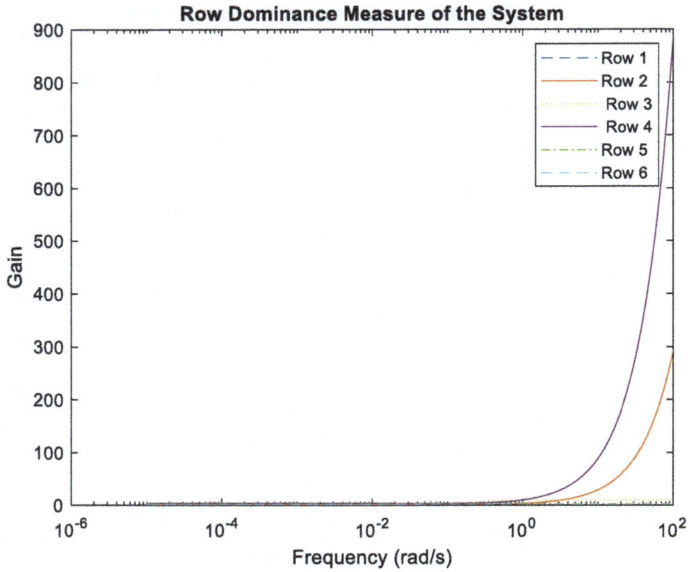

Fig. 4.21 Row dominance for Edmunds scaling at different frequency

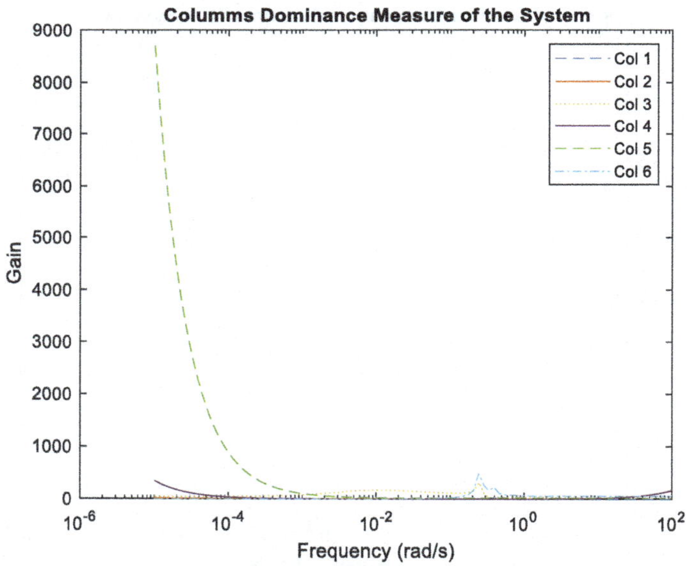

Fig. 4.22 Column dominance for Edmunds scaling at different frequency

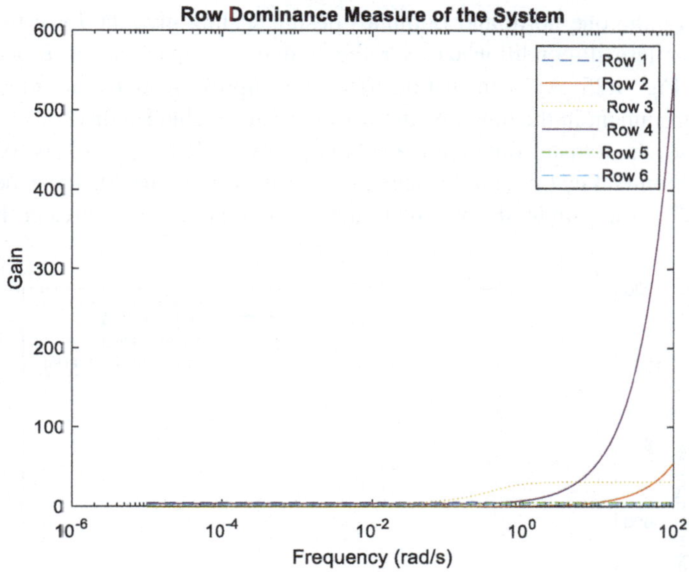

Fig. 4.23 Row dominance for Perron-Frobenius scaling at different frequency

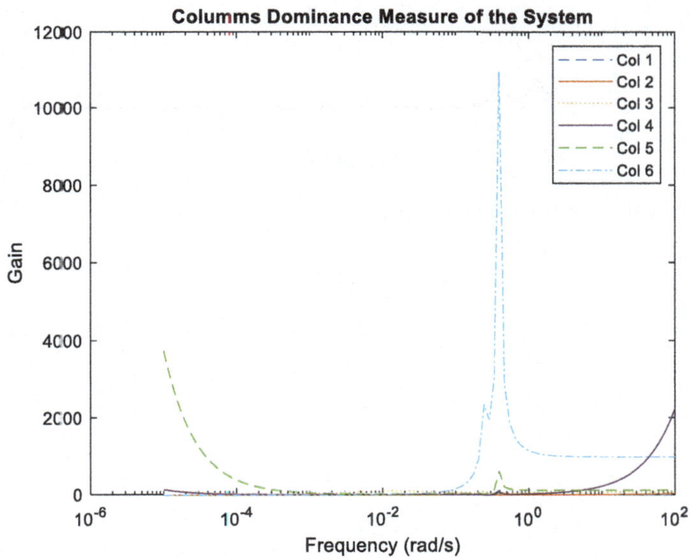

Fig. 4.24 Column dominance for Perron-Frobenius scaling at different frequency

However, the one-norm and Edmunds design scaling give the lowest condition number (or less ill-conditioned) over the frequencies than Perron-Frobenius, as shown in Fig. 4.25. As seen in Fig. 4.26, the improvement in the interaction is not very prominent in the direct Nyquist array with Gershgorin discs.

However, the transfer function matrix (TFM) of the ROV is unlikely to be made diagonal dominant by this simple diagonal constant scaling matrix. It can be verified in Fig. 4.27 as the gain in dB over the frequencies of interest is all greater than 6 dB.

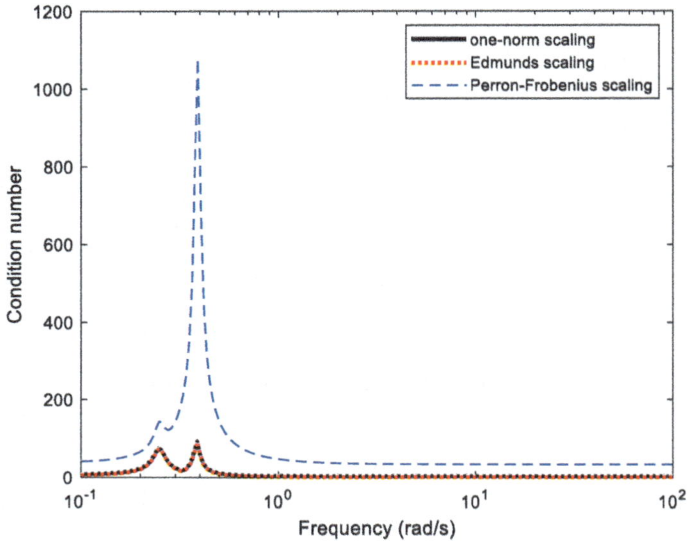

Fig. 4.25 Condition number at different frequency

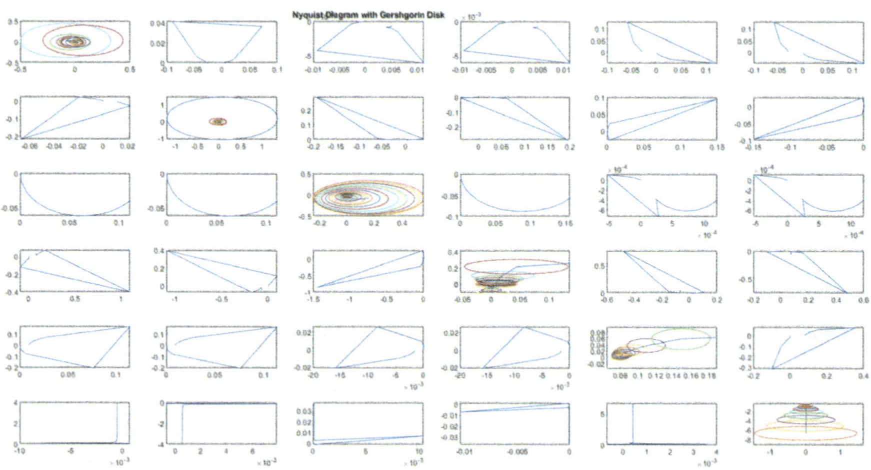

Fig. 4.26 Direct Nyquist array with Gershgorin discs after Edmunds scaling

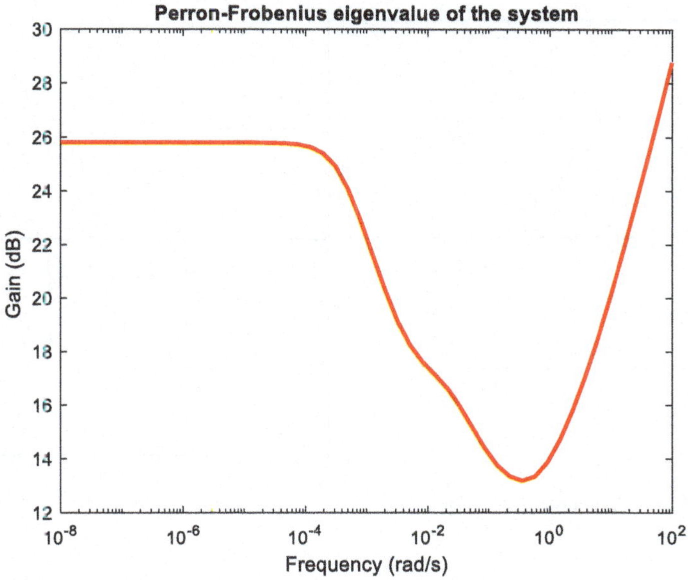

Fig. 4.27 Perron-Frobenius eigenvalues of ROV system

MATLAB Script

```
rrr=50; ccc=2; p1=1;
table_1=cell(rrr,ccc);
s=logspace(-8,2);

for n=1:rrr
  G=abs(evalfr(sys_rov,s(n)*i));
  Gdiag=diag(diag(G));
  invGdiag=minv(Gdiag);
  GG=mmult(invGdiag,G);
  perGG=eig(GG);
 [pfval,vector_left,vector_right,gs]=speron(GG);
 vec_left1(n)=vector_left(1,1);
 vec_left2(n)=vector_left(2,2);
 vec_left3(n)=vector_left(3,3);
 vec_left4(n)=vector_left(4,4);
 vec_left5(n)=vector_left(5,5);
 vec_left6(n)=vector_left(6,6);
 maxperGG(n)=20*log10(max(perGG));
  table_1{p1,1}=s(n);          % create cell to store maxperGG
  table_1{p1,2}=maxperGG(n);
  p1=p1+1;
end

figure();
semilogx(s,maxperGG,'r');
title('Perron-Frobenius eigenvalue of the system');
xlabel('Frequency (rad/s)');
ylabel('Gain (dB)');
```

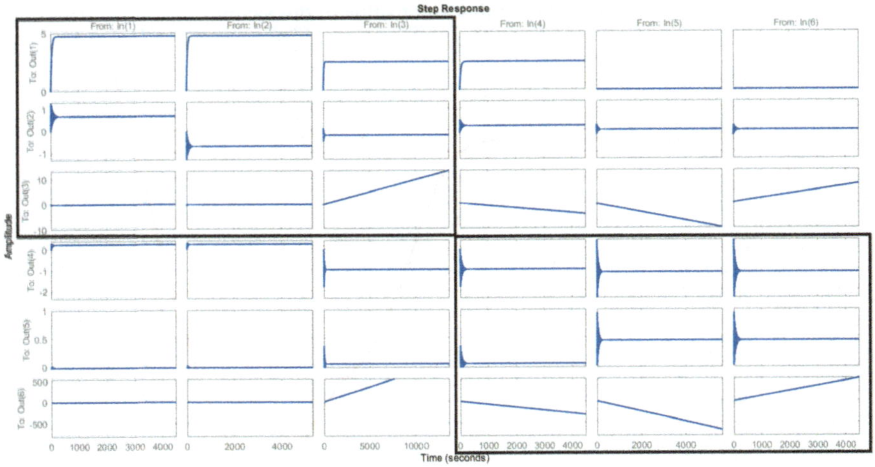

Fig. 4.28 Mostly decoupled step response

The Edmunds scaling using the non-unity permutation matrices is chosen. It helps to numerically condition and reduces the interactions between the input and the output interaction, as it is not always possible to have a clear diagonal dominance at each row or column. As shown in Fig. 4.28, the ROV system can be grouped into two blocks of 3×3 subsystems. Instead of looking at each element in each row (column) for the row (column) diagonal dominance, a group of elements in a block can be tested for diagonal dominance. This is known as the block diagonal dominance measure of the TFM:

$$\mathbf{G}_{\mathbf{ij}}^{\mathbf{k}\times\mathbf{k}}(\mathbf{s}) = \begin{bmatrix} \mathbf{G}_{\mathbf{ii}}^{3\times 3} & \mathbf{G}_{\mathbf{ij}}^{3\times 3} \\ \mathbf{G}_{\mathbf{ji}}^{3\times 3} & \mathbf{G}_{\mathbf{jj}}^{3\times 3} \end{bmatrix} \tag{4.35}$$

where superscript 3×3 and subscripts i, j refer to the size of the partitioned block and elements in the block, respectively. The matrix $\mathbf{G}(s)$ is considered to be block diagonal dominant if the gain of the diagonal blocks dominates the gain of the off-diagonal blocks. To test the block diagonal dominance of $\mathbf{G}(s)$, the gains of the sub-blocks of $\mathbf{G_{ii}}$ are determined as follows:

$$\begin{aligned} \underline{\sigma}(\mathbf{G_{ii}}) &= \min_{x\neq 0} \frac{\|\mathbf{G_{ii}x}\|}{\|\mathbf{x}\|} \\ &= \text{smallest singular value of } \mathbf{G_{ii}} \end{aligned} \tag{4.36}$$

The pre-compensator matrices are as follows:

$$\mathbf{S}_{\text{pre}} = \begin{bmatrix} 0 & 0 & 0 & 2.732 & 0 & 0 \\ 0 & 0 & 2.736 & 0 & 0 & 0 \\ 0 & 3.752 & 0 & 0 & 0 & 0 \\ 3.752 & 0 & 0 & 0 & 0 & 0 \\ 0 & 0 & 0 & 0 & 2.773 & 0 \\ 0 & 0 & 0 & 0 & 0 & 2.772 \end{bmatrix} \in \Re^{6\times 6} \tag{4.37}$$

The post-compensator matrices are as follows:

$$\mathbf{S}_{\text{post}} = \begin{bmatrix} 0 & 0 & 7.195 & 0 & 0 & 0 \\ 0 & 6.763 & 0 & 0 & 0 & 0 \\ 0 & 0 & 0 & 0.948 & 0 & 0 \\ 13.466 & 0 & 0 & 0 & 0 & 0 \\ 0 & 0 & 0 & 0 & 2.019 & 0 \\ 0 & 0 & 0 & 0 & 0 & 0.644 \end{bmatrix} \in \Re^{6\times 6} \tag{4.38}$$

By using the Edmunds scaling, the following I/O pairs for the ROV system were obtained: input 1—heave, input 2—surge, input 3—sway, input 4—roll, input 5—pitch, and input 6—yaw.

To verify that the ROV system is now block-diagonal dominant, Fig. 4.29 shows the block dominance measure of the system before and after the Edmunds scaling using $\underline{\sigma}(\mathbf{G})$. As shown, the ROV model becomes more block-diagonal dominance. The block dominance measure of TFM shows a decrease from 3.7×10^{-5} to 6.5×10^{-8}. It implies that the TFM of the ROV is now decoupled into two blocks of each matrix of size 3×3.

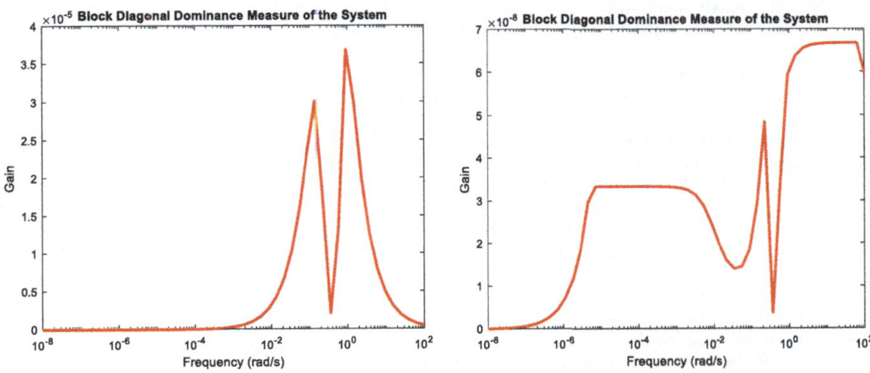

Fig. 4.29 Block-diagonal dominance measure before (left) and after (right) Edmunds scaling

MATLAB Script

```
s=logspace(-8,2);
clear i table_domin table_domin2 table_domin3

% initialise the table/cell
rro=50;
cco=3;
table_domin=cell(rro,cco);
table_domin2=cell(rro,3);
table_domin3=cell(rro,3);
pp1=1;

for n=1:rro
   % min svd of diagonal block / max svd of off-diagonal block--> Block diagonal dominance measure
  G_sv=evalfr(G_tot,s(n)*i);
  G33_sv1=G_sv(1:3,1:3);
  G33_sv2=G_sv(4:6,4:6);
  diag_block=[G33_sv1     zeros(3) ;     zeros(3)   G33_sv2];
   off_diag_block=G_sv-diag_block;

 sv_min=min(svd(abs(diag_block)));
 sv_max=max(svd(abs(off_diag_block)));
 ratio_sv(n)=sv_min/(sv_max);

  % min svd of diagonal block / max svd off- diagonal block--> Block diagonal row dominance measure
 svsum_off_diag_row_1=sum(abs(off_diag_block(1,:)))+sum(abs(off_diag_block(2,:)))+ ...
 sum(abs(off_diag_block(3,:)));
  svsum_off_diag_row_2=sum(abs(off_diag_block(4,:)))+sum(abs(off_diag_block(5,:)))+...
  sum(abs(off_diag_block(6,:))) ;

 sv_row1_ratio(n)=min(svd(G33_sv1))/max(svd(svsum_off_diag_row_1));
 sv_row2_ratio(n)=min(svd(G33_sv2))/max(svd(svsum_off_diag_row_2));

 % min svd of diagonal block / max svd off- diagonal block--> Block diagonal col dominance measure
 svsum_off_diag_col_1=sum(abs(off_diag_block(:,1)))+sum(abs(off_diag_block(:,2)))+ ...
 sum(abs(off_diag_block(:,3)));
 svsum_off_diag_col_2=sum(abs(off_diag_block(:,4)))+ sum(abs(off_diag_block(:,5)))+...
 sum(abs(off_diag_block(:,6)));

 sv_col1_ratio(n)=min(svd(G33_sv1))/max(svd(svsum_off_diag_col_1));
 sv_col2_ratio(n)=min(svd(G33_sv2))/max(svd(svsum_off_diag_col_2));

  % fundamental diagonal dominance
 spectral_radius(n)=max(abs(eig(off_diag_block*inv(diag_block))));

 % col dominance of G(s)----> max of the col
 B_invD=off_diag_block*inv(diag_block);
 norm_one_BinvD(n)=norm(B_invD,1);

 % row dominance of G(s)----> max of the row
 norm_inf_BinvD(n)=norm(B_invD,inf);

 % rows dominance ratios
 sum_off_diag_row_1=sum(abs(off_diag_block(1,:)))+sum(abs(off_diag_block(2,:)))+ ...
 sum(abs(off_diag_block(3,:)));
 sum_off_diag_row_2=sum(abs(off_diag_block(4,:)))+ ...
 sum(abs(off_diag_block(5,:)))+sum(abs(off_diag_block(6,:)));
 ratio_row1(n)=(abs(sum(sum(G33_sv1))));
 ratio_row2(n)=(abs(sum(sum(G33_sv2))));
 tot_sum_row=sum_off_diag_row_1+sum_off_diag_row_2;
```

```
ratio_row(n)=(ratio_row1(n)+ratio_row2(n) )/tot_sum_row;

% columns dominance ratios
sum_off_diag_col_1=sum(abs(off_diag_block(:,1)))+ sum(abs(off_diag_block(:,2)))+ ...
sum(abs(off_diag_block(:,3)));
sum_off_diag_col_2=sum(abs(off_diag_block(:,4)))+...
sum(abs(off_diag_block(:,5)))+sum(abs(off_diag_block(:,6)));
ratio_col1(n)=(abs(sum(sum(G33_sv1))));
ratio_col2(n)=(abs(sum(sum(G33_sv2))));
tot_sum_col=sum_off_diag_col_1+sum_off_diag_col_2;
ratio_col(n)=(ratio_col1(n)+ratio_col2(n) )/tot_sum_col;

% create cell to store ratio
 table_domin{pp1,1}=s(n);
 table_domin{pp1,2}=ratio_row1(n);
 table_domin{pp1,3}=ratio_row2(n);
 table_domin2{pp1,1}=s(n);
 table_domin2{pp1,2}=ratio_col1(n);
 table_domin2{pp1,3}=ratio_col2(n);
 table_domin3{pp1,1}=s(n);
 table_domin3{pp1,2}=norm_inf_BinvD(n);
 table_domin3{pp1,3}=norm_one_BinvD(n);
 pp1=pp1-1;
end
```

The proposed Edmunds scaling algorithm on the TFM sets all the row and column sums to unity using simple iterative steps. The algorithm equalizes and maximizes the geometric mean of the row dominance and geometric mean of the column dominance for given inputs and outputs. It improves the numerical stability in MPC system design where the solution of the algebraic Riccati equation is needed. Also, the design scaling makes the individual loops more independent by balancing the interactions between the loops. As a result, the scaled state-space model of ROV becomes:

$$\dot{\mathbf{x}} = \mathbf{A}_{11}\mathbf{x} + \mathbf{B}_1\mathbf{S}_{\mathbf{pre}}\bar{\mathbf{u}} \tag{4.39}$$

$$\mathbf{y} = \mathbf{S}_{\mathbf{post}}\mathbf{C}_{11}\mathbf{x} \tag{4.40}$$

where $\mathbf{S}_{\mathbf{pre}} \in \Re^{6\times 6}$ is the pre-compensator and $\mathbf{S}_{\mathbf{post}} \in \Re^{6\times 6}$ is the post-compensator using the Edmunds design scaling method.

4.6.2 Decentralized Model Predictive Control

In the research literature, model predictive control (MPC) is formulated in state-space. The MPC is obtained by minimizing the cost function by application of the quadratic programming algorithms and the relevant references cited in this text. The model of the ROV is described by the linear difference equations in (4.39) and (4.40). As the input $\bar{\mathbf{u}}$ cannot affect the output $\mathbf{y}$ at the same time, $\mathbf{D} = 0$ is assumed. The following auxiliary variables are defined to include the state-space model in an augmented state-space model.

$$\mathbf{z}(t) = \dot{\mathbf{x}}(t) \tag{4.41}$$

$$\mathbf{y}(t) = \mathbf{C}_{11}\mathbf{x}(t) \tag{4.42}$$

With a new state vector $\mathbf{x_a}(t) = [\mathbf{z}(t)^{\mathrm{T}}\,\mathbf{y}(t)^{\mathrm{T}}]^{\mathrm{T}}$, the augmented state-space model is defined as:

$$\begin{bmatrix} \dot{\mathbf{z}}(t) \\ \dot{\mathbf{y}}(t) \end{bmatrix} = \overbrace{\begin{bmatrix} \mathbf{A}_{11} & \mathbf{0}_{n\times m} \\ \mathbf{S_{post}}\mathbf{C}_{11} & \mathbf{0}_{m\times m} \end{bmatrix}}^{\mathbf{A}} \begin{bmatrix} \mathbf{z}(t) \\ \mathbf{y}(t) \end{bmatrix} + \overbrace{\begin{bmatrix} \mathbf{B}_1\mathbf{S_{pre}} \\ \mathbf{0}_{m\times m} \end{bmatrix}}^{\mathbf{B}} \dot{\bar{\mathbf{u}}}(t) \tag{4.43}$$

$$\mathbf{y}(t) = \overbrace{\begin{bmatrix} \mathbf{0}_{m\times n} & \mathbf{I}_{m\times m} \end{bmatrix}}^{\mathbf{C}} \begin{bmatrix} \mathbf{z}(t) \\ \mathbf{y}(t) \end{bmatrix} \tag{4.44}$$

where $\mathbf{I}_{m\times m}$ is the identity matrix $m \times m$; $\mathbf{0}_{m\times m}$ and $\mathbf{0}_{m\times n}$ are zero matrices of dimension $m \times m$ and $m \times n$, respectively.

The information of the state vector at a time t_i is written as $\mathbf{x}(t_i)$. The state feedback control with gain matrix $\mathbf{K}$ is valid for $0 \le \tau \le T_p$ where τ is the time variable within the window from t_i to prediction horizon T_p. The derivative of the control within this window is therefore written as:

$$\dot{\bar{\mathbf{u}}}(\tau) = -\mathbf{K}\mathbf{x}(\tau) \tag{4.45}$$

The predicted future state at time τ

$$\mathbf{x}(t_i + \tau | t_i) = e^{\mathbf{A}\tau}\mathbf{x}(t_i) + \mathbf{\Phi}(\tau)^{\mathrm{T}}\boldsymbol{\eta} \tag{4.46}$$

where

$$\mathbf{\Phi}(\tau)^{\mathrm{T}} = \int_0^{\tau} e^{\mathbf{A}(\tau-\gamma)}[\,\mathbf{B_1L_1}(\gamma)^{\mathrm{T}}\ \mathbf{B_2L_2}(\gamma)^{\mathrm{T}} \ldots \mathbf{B_mL_m}(\gamma)^{\mathrm{T}}\,]d\gamma;\ \boldsymbol{\eta}^{\mathrm{T}} = [\,\boldsymbol{\eta}_1^{\mathrm{T}}\ \boldsymbol{\eta}_2^{\mathrm{T}} \ldots \boldsymbol{\eta}_{\mathrm{m}}^{\mathrm{T}}\,] \tag{4.47}$$

$$\mathbf{L_i}(\tau)^{\mathrm{T}} = [\,l_1^i(\tau)\ l_2^i(\tau) \ldots l_{N_p}^i(\tau)\,] = e^{\mathbf{A_p}\tau}\mathbf{L}(0) \tag{4.48}$$

$$\mathbf{A_p} = \begin{bmatrix} -\rho_A & 0 & \ldots & 0 \\ -2\rho_A & -\rho_A & \ldots & 0 \\ \vdots & \vdots & \ldots & \vdots \\ -2\rho_A & \cdots & \cdots & -2\rho_A - \rho_A \end{bmatrix} \tag{4.49}$$

$$\mathbf{L}(0) = \sqrt{2\rho_A}\big[\,1\ 1 \ldots 1\,\big]^{\mathrm{T}} \tag{4.50}$$

where ρ_A is the time scaling factor and N_p is the Laguerre parameter for each input.

The predicted output at a time τ can be computed as follows:

$$\mathbf{y}(t_i + \tau | t_i) = \mathbf{C}[e^{\mathbf{A}\tau}\mathbf{x}(t_i) + \mathbf{\Phi}(\tau)^{\mathrm{T}}\boldsymbol{\eta}] \tag{4.51}$$

where

$$\boldsymbol{\Omega} = \int_0^{T_p} [\mathbf{\Phi}(t_i)\mathbf{Q}\Phi(t_i)^{\mathrm{T}} + \mathbf{R}]d\tau;\ \ \mathbf{Q}, \mathbf{R} \geq 0 \tag{4.52}$$

$$\boldsymbol{\Psi} = \int_0^{T_p} \mathbf{\Phi}(t_i)\mathbf{Q}e^{\mathbf{A}\tau}d\tau \tag{4.53}$$

The optimal $\boldsymbol{\eta}$ becomes

$$\boldsymbol{\eta} = -\boldsymbol{\Omega}^{-1}\Psi\mathbf{x}(t_i) \tag{4.54}$$

that minimizes $\mathbf{J}$ as

$$\mathbf{J}_{\min} = \mathbf{x}(t_i)^{\mathrm{T}}\left[\int_0^{T_p} e^{\mathbf{A}^{\mathrm{T}}\tau}\mathbf{Q}e^{\mathbf{A}\tau}d\tau - \boldsymbol{\Psi}^{\mathrm{T}}\boldsymbol{\Omega}^{-1}\boldsymbol{\Psi}\right]\mathbf{x}(t_i) \tag{4.55}$$

The optimal control signal is written as:

$$\begin{aligned}\dot{\mathbf{u}}(t) &= \begin{bmatrix} L_1(0)^{\mathrm{T}} & 0_2 & \dots & 0_m \\ 0_1 & L_2(0)^{\mathrm{T}} & \dots & 0_m \\ \vdots & \vdots & \dots & \vdots \\ 0_1 & 0_2 & \dots & L_m(0)^{\mathrm{T}} \end{bmatrix}\begin{bmatrix} \eta_1 \\ \eta_2 \\ \vdots \\ \eta_m \end{bmatrix} \\ &= -\mathbf{L}(\tau)^{\mathrm{T}}\boldsymbol{\Omega}^{-1}\Psi\mathbf{x}(t) \\ &= -\mathbf{K}_{\mathbf{mpc}}\mathbf{x}(t) \end{aligned} \tag{4.56}$$

As the first components are the derivatives of the input state $\mathbf{x}(t)$, with a set-point signal $\mathbf{r}(t)$, the error signal $\mathbf{y}(t) - \mathbf{r}(t)$ is the second part of the feedback element. Thus, the state-feedback becomes:

$$\dot{\mathbf{u}}(t) = -\overbrace{\begin{bmatrix} \mathbf{K}_{\mathbf{x}}\ \mathbf{K}_{\mathbf{y}} \end{bmatrix}}^{\mathbf{K}_{\mathbf{mpc}}}\begin{bmatrix} \dot{\mathbf{x}}(t) \\ \mathbf{y}(t) - \mathbf{r}(t) \end{bmatrix} \tag{4.57}$$

The control signal computation is carried out by simply approximating the derivative $\dot{\bar{\mathbf{u}}}(t)$ with

$$\dot{\bar{\mathbf{u}}}(t_i) \approx \frac{\bar{\mathbf{u}}(t_i) - \bar{\mathbf{u}}(t_i - \Delta t)}{\Delta t} \Rightarrow \bar{\mathbf{u}}(t_i) = \bar{\mathbf{u}}(t_i - \Delta t) + \dot{\bar{\mathbf{u}}}(t_i)\Delta t \tag{4.58}$$

The actual control signal $\bar{\mathbf{u}}(t)$ will update itself repeatedly from its past value.

$$\begin{aligned} \bar{\mathbf{u}}(t_i) &= \bar{\mathbf{u}}_{\mathbf{actual}}(t_i - \Delta t) + \dot{\bar{\mathbf{u}}}(t_i)\Delta t \\ &= \bar{\mathbf{u}}_{\mathbf{actual}}(t_i - \Delta t) + \begin{bmatrix} L_1(0)^{\mathrm{T}} & 0_2 & \ldots & 0_m \\ 0_1 & L_2(0)^{\mathrm{T}} & \ldots & 0_m \\ \vdots & \vdots & \ldots & \vdots \\ 0_1 & 0_2 & \cdots & L_m(0)^{\mathrm{T}} \end{bmatrix} \begin{bmatrix} \eta_1 \\ \eta_2 \\ \vdots \\ \eta_m \end{bmatrix} \Delta t \end{aligned} \tag{4.59}$$

The optimal solution with the exponentially weighted cost function is identical to the original optimal solution without exponential weighting when the prediction horizon is sufficiently large. However, without modification on the pair of weight matrices $\mathbf{Q}$ and $\mathbf{R}$, the solution does not guarantee exponential decay of the original variable, $\mathbf{x}(t_i + \tau | t_i)$ within the horizon. The weight matrix $\mathbf{Q}$ becomes $\mathbf{Q}_\alpha = \mathbf{Q} + 2(\alpha + \beta)\mathbf{P}$; $\alpha, \beta > 0, \mathbf{R} > 0$. $\mathbf{P}$ is the solution of the steady-state Riccati equation that remains unchanged. The optimal control $\dot{\bar{\mathbf{u}}}(\tau)$ is thus obtained. The proposed approach is numerically sound as it allows the use of a sufficiently large prediction horizon using $\mathbf{A}_\alpha = \mathbf{A} - \alpha\mathbf{I}$ to guarantee asymptotical stability.

The control amplitude constraints for the state-space model at a time t can be written in the form where $\bar{\mathbf{u}}_{\mathbf{min}}$ and $\bar{\mathbf{u}}_{\mathbf{max}}$ are vectors containing the required lower and upper limits of the control amplitude.

$$\bar{\mathbf{u}}_{\mathbf{min}} \leq \bar{\mathbf{u}}(t) \leq \bar{\mathbf{u}}_{\mathbf{max}} \tag{4.60}$$

The constraints on $\dot{\bar{\mathbf{u}}}(t)$ take the form

$$\dot{\bar{\mathbf{u}}}_{\mathbf{min}} \leq \dot{\bar{\mathbf{u}}}(t) \leq \dot{\bar{\mathbf{u}}}_{\mathbf{max}} \tag{4.61}$$

where $\dot{\bar{\mathbf{u}}}_{\mathbf{min}}$ and $\dot{\bar{\mathbf{u}}}_{\mathbf{max}}$ correspond to the minimum and maximum changes in the thruster rate, respectively. The constraints are applied to the system inputs on both the absolute limits and the speed of change. The rate of the thrust input is chosen to be lower than the maximum rate (this is an estimate as the thrust is proportional to square of its speed). If the controller requires a significant change of thrust beyond the thruster specification, then the thruster dynamics will perform like a low-pass filter to improve the controller performance. The rate of change of setpoints for each thruster is set to ±0.001 N/s. The maximum (minimum) absolute thrust value is limited to 100 N (−100 N), corresponding to a thruster speed of approximately 1000 r/min.

The implementation of MPC law also requires direct measurement of all elements $[\dot{\mathbf{x}}(t)\,\mathbf{y}(t) - \mathbf{r}(t)]^{\mathrm{T}}$ at each time instant. The state vector component $\dot{\mathbf{x}}(t)$ can be estimated. However, there is noise corruption in $\mathbf{x}(t)$, and the computed difference $\dot{\mathbf{x}}(t)$ will amplify the noise. Thus, an observer is used to estimate $\dot{\mathbf{x}}(t)$ as

$$\dot{\hat{\mathbf{x}}}(t) = \mathbf{A}\hat{\mathbf{x}}(t) + \mathbf{B}\dot{\mathbf{u}}(t) + \mathbf{K}_{\mathbf{ob}}[\mathbf{y}(t) - \mathbf{C}\hat{\mathbf{x}}(t)] \tag{4.62}$$

where $\hat{\mathbf{x}}(t)$ is the estimate of $\mathbf{x}(t)$ and $\mathbf{K}_{\mathbf{ob}}$ is the observer gain matrix. $\mathbf{K}_{\mathbf{ob}}$ is designed using discrete-linear quadratic regulator theory with weighting matrices $\mathbf{Q}_{\mathbf{ob}}$ and $\mathbf{R}_{\mathbf{ob}}$. $\mathbf{K}_{\mathbf{ob}}$ is selected such that the closed-loop observer error system matrix is stable. The effects of measurement noise in $\mathbf{y}(t)$ can be reduced by choosing a large value of $\mathbf{R}_{\mathbf{ob}}$. The initial state vector of the observer is set to 0 for output vector $\mathbf{y}(t) = 0$. The block diagram for simulating the decentralized model predictive control can be seen in Fig. 4.30. The desired velocities are set at $u = 0.1$ m/s and $v = 0.1$ m/s. The observer block can be modeled as shown in Fig. 4.31.

In summary, the steps used to design the decentralized MPC with Edmunds scaling are as follows:

(1) Determine the non-unity permutation matrices ($\mathbf{S}^{\mathbf{E}}_{\mathbf{post}}$ and $\mathbf{S}^{\mathbf{E}}_{\mathbf{pre}}$) of Edmunds scaling to achieve the least interaction and block-diagonal dominance on the transfer function Matrix;

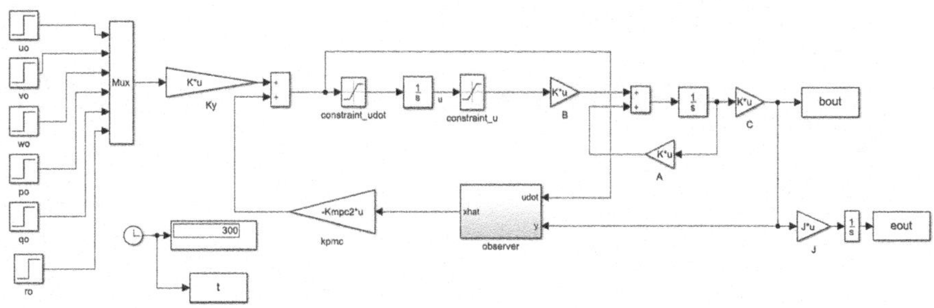

Fig. 4.30 Decentralized model predictive control for ROV

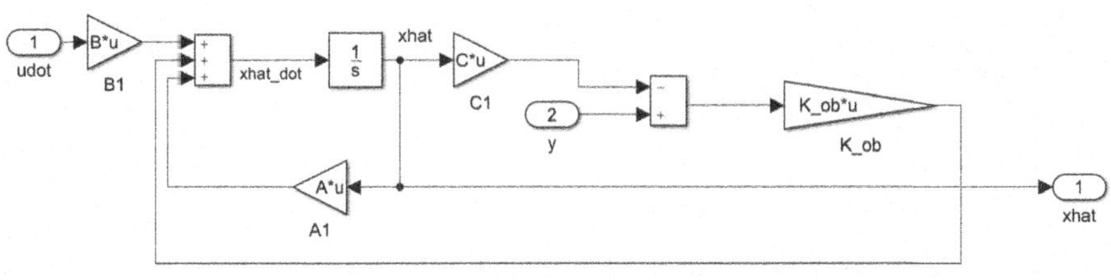

Fig. 4.31 Observer for decentralized model predictive control of ROV

MATLAB SCRIPT
```
% minimization of interaction-scaling method- Edmund scaling (gs=post*g*pre)
w1=0.38;
Gw1=evalfr(sys_rov,w1);
[pre1e,post1e,gs1e,blocks1e,block1e]=normalise(Gw1);
sys_tot_e=post1e*sys_rov*pre1e;
[Ap, Bp,Cp, Dp]=ssdata(sys_tot_e)
```

(2) Form the augmented matrix for MPC using Edmunds scaling;

MATLAB SCRIPT
```
% Augmented matrix
[m1,n1]=size(Cp)
[n1,n_in]=size(Bp)
A=zeros(n1+m1,n1+m1);
A(1:n1,1:n1)=Ap;
A(n1+1:n1+m1,1:n1)=Cp;
B=zeros(n1+m1,n_in);
B(1:n1,:)=Bp;
C=zeros(m1,n1+m1);
C(:,n1+1:n1+m1)=eye(m1,m1);
```

(3) Set parameters for continuous MPC: time scaling parameter (ρ_A), number of Laguerre parameters (N_p), prediction horizon (T_p), and plane shifting parameters on matrix $\mathbf{A}(\alpha, \beta > 0)$;

MATLAB SCRIPT
```
p1=15;% scaling parameter- higher better-rate of increase
p2=15;

N1=6;% no of Laguerre parameters for each input
N2=6;

Tp1=25;   % prediction horizon - 20
Tp2=25;

p=[p1*ones(1,8) p2*ones(1,8)];            % scaling parameter
N=[N1*ones(1,8)  N2*ones(1,8) ];          % no of Laguerre parameters for each input
Tp=[Tp1*ones(1,8)  Tp2*ones(1,8) ];       % prediction horizon
```

(4) Solve Riccati equation for $\mathbf{P}$ using $\mathbf{Q}_\alpha = \mathbf{Q} + 2(\alpha + \beta)\mathbf{P} > 0, \mathbf{R} > 0$ and $\mathbf{A}_\alpha = \mathbf{A} - \alpha\mathbf{I}$;

MATLAB SCRIPT
```
Q=C'*C;
R=eye(m1,m1)
n=n1+m1;
beta=0.01;
[K,P,E]=lqr(A+beta*eye(n,n),B,Q,R);
```

(5) Solve MPC for $\mathbf{\Omega}$ and $\mathbf{\Psi}$ using (4.52) and (4.53), respectively;

MATLAB SCRIPT
```
alpha=8;    % must be greater than zero to be stable
 Q_alpha=Q+2*(alpha+beta)*P;
 A_alpha=A-alpha*eye(n,n);
 [Omega2,Psi2]=cmpc(A_alpha,B,p,N,Tp,Q_alpha,R);
```

```
MATLAB SCRIPT
function [Omega,Psi]=cmpc(A,B,p,N,Tp,Q,R);
[n,n_in]= size(B);
tau_del=0.001/max(p);
Tpm=max(Tp);
tau=0:tau_del:Tpm;
Np=length(tau);
N_pa=sum(N);
Omega=zeros(N_pa,N_pa);
Psi=zeros(N_pa,n);
S_in=zeros(n,N_pa);

% RL is the weight matrix corresponding to the weight on the coefficient
R_L=eye(N_pa,N_pa);
kk=1;
for i=1:n_in
R_L(kk:kk-1+N(i),kk:kk-1+N(i))= R(i,i)*R_L(kk:kk-1+N(i),kk:kk-1+N(i));
kk=kk+N(i);
end
 [Al,L0]=lagc(p(1),N(1));
 Eae=expm(A*tau_del);
 Eap=expm(Al*tau_del);
 L=Eap*L0;
 Y=-B(:,1)*L'+Eae*B(:,1)*L0';
 X=Iint(A,p(1),Y);
 S_in(:,1:N(1))=X;
 In_s=1;
 for jj=2:n_in;
 [Al,L0]=lagc(p(jj),N(jj));
 Eap=expm(Al*tau_del);
 L=Eap*L0;
 Y=-B(:,jj)*L'+Eae*B(:,jj)*L0';
 X=Iint(A,p(jj),Y);
 In_s=N(jj-1)+In_s;
 In_e=In_s+N(jj)-1;
 S_in(:,In_s:In_e)=X;
 end
 S_sum=S_in;

 for i=2:Np-1;
 kk=1;
 [Al,L0]=lagc(p(kk),N(kk));
 Eap=expm(Al*tau_del);
 S_sum(:,1:N(kk))=Eae*S_sum(:,1:N(kk))+S_in(:,1:N(kk))*(Eap^(i-1))';
 In_s=1;
 for kk=2:n_in;
 [Al,L0]=lagc(p(kk),N(kk));
 Eap=expm(Al*tau_del);
 In_s=N(kk-1)+In_s;
 In_e=In_s+N(kk)-1;
 S_sum(:,In_s:In_e)=Eae*S_sum(:,In_s:In_e)+S_in(:,In_s:In_e)*(Eap^(i-1))';
 end
 phi=S_sum;
 Omega=Omega+phi'*Q*phi;
 Psi=Psi+phi'*Q*Eae^i;
 end

 Omega=Omega*tau_del+R_L;
 Psi=Psi*tau_del;
```

(6) Compute observer gain matrix, $\mathbf{K_{ob}}$ with weighting matrices $\mathbf{Q_{ob}}$ and $\mathbf{R_{ob}}$.

```
MATLAB SCRIPT
Q1=eye(n,n);
R1=13*eye(m1,m1);
K_ob=lqr(A',C',Q1,R1)';
```

(7) Define constraint $\mathbf{L}(\tau)$ using (4.48) and input constraints on thrusters: $\bar{\mathbf{u}}_{\mathbf{min}} \leq \bar{\mathbf{u}}(t) \leq \bar{\mathbf{u}}_{\mathbf{max}}$;

```
MATLAB Script
xm=zeros(n1,1);
u=zeros(n_in,1);
y=zeros(m1,1);
h=0.01; %0.009 to 0.01;
N_sim=10*3500;
sp1=ones(n1,N_sim)*0;                        % set to zero
sp2=[ones(2,N_sim)*0.1;zeros(4,N_sim) ];  % reference input, r  for all  6 inputs
sp=[sp1;sp2];                                 % set-point signal

[Md,Lzerot]=Mder(p,N,n_in,0.1);
[Mu,Mu1]=Mucon(p,N,n_in,h,0.1);
M=[Mu;-Mu;Lzerot;-Lzerot];

u_max=[100*ones(m1,1)];
u_min=[-100*ones(m1,1)];
Deltau_max=[0.001*ones(m1,1)];
Deltau_min=[-0.001*ones(m1,1)];
```

```
MATLAB Script
function [M_dv,Lzerot]=Mder(p,N,n_in,tau)
N_pa=sum(N);
k0=1;
[Al,L0]=lagc(p(k0),N(k0));
L_t=zeros(n_in,N_pa);
L_t(1,1:N(1))=L0';
cc=N(1);
for k0=2:n_in;
[Al,L0]=lagc(p(k0),N(k0));
L_t(k0,cc+1:cc+N(k0))=L0';
cc=cc+N(k0);
end
Lzerot=L_t;
k0=1;
[Al,L0]=lagc(p(k0),N(k0));
L1=expm(Al*tau)*L0;
L_t1=zeros(n_in,N_pa);
L_t1(1,1:N(1))=L1';
cc=N(1);
for k0=2:n_in;
[Al,L0]=lagc(p(k0),N(k0));
L1=expm(Al*tau)*L0;
L_t1(k0,cc+1:cc+N(k0))=L1';
cc=cc+N(k0);
end
M_dv=[L_t1];
```

MATLAB Script
```
function [Mu,Mu1]=Mucon(p,N,n_in,delta_t,tau)
N_pa=sum(N);
k0=1;
[Al,L0]=lagc(p(k0),N(k0));
L_t=zeros(n_in,N_pa);
L_t(1,1:N(1))=L0';
cc=N(1);
for k0=2:n_in;
[Al,L0]=lagc(p(k0),N(k0));
L_t(k0,cc+1:cc+N(k0))=L0';
cc=cc+N(k0);
end

% constraints on second sample
k0=1;
[Al,L0]=lagc(p(k0),N(k0));
L1=expm(Al*tau)*L0;
L_t1=zeros(n_in,N_pa);
L_t1(1,1:N(1))=(L1'-L0')*inv(Al')+L0'*delta_t;
cc=N(1);

for k0=2:n_in;
[Al,L0]=lagc(p(k0),N(k0));
L1=expm(Al*tau)*L0;
L_t1(k0,cc+1:cc+N(k0))=(L1'-L0')*inv(Al')+L0'*delta_t;
cc=cc+N(k0);
end
Mu=[L_t*delta_t];
Mu1=[L_t1];
```

(8) Compute the derivative state-feedback gain for $\dot{\bar{\mathbf{u}}}(t) = -\mathbf{K_{mpc}}\mathbf{x}(t)$ where $\mathbf{K_{mpc}} = \mathbf{L}(\tau)^T\mathbf{\Omega}^{-1}\mathbf{\Psi}$ using (4.56);

MATLAB Script
```
Kmpc2=Lzerot*(Omega2\Psi2)
Acl=A_alpha-B*Kmpc2;
eig(Acl)
```

(9) Compute the state feedback gain $\bar{\mathbf{u}}(t)$;

MATLAB Script
```
[u1,y1,udot1,t1]=cssimucon1(xm,u,y,sp,Ap,Bp,Cp,A_alpha,B,C,N_sim,Omega2,Psi2,K_ob,Lzerot,
h,M,u_max,u_min,Deltau_max,Deltau_min);
```

MATLAB Script

```
function eta=QPhild(H,f,A_cons,b)
% E=H;
% F=f;
% M=A_cons;
% gamma=b;
% eta =x
[n1,m1]=size(A_cons);
eta=-H\f;
kk=0;
for i=1:n1
if (A_cons(i,:)*eta>b(i)) kk=kk+1;
else
kk=kk+0;
end
end
if (kk==0) return; end

P=A_cons*(H\A_cons');
d=(A_cons*(H\f)+b);
[n,m]=size(d);
x_ini=zeros(n,m);
lambda=x_ini;
al=10;
for km=1:38

lambda_p=lambda;
for i=1:n
w= P(i,:)*lambda-P(i,i)*lambda(i,1);
w=w+d(i,1);
la=-w/P(i,i);
lambda(i,1)=max(0,la);
end
al=(lambda-lambda_p)'*(lambda-lambda_p);
if (al<10e-8); break; end
end

eta=-inv(H)*f -inv(H)*A_cons'*lambda;
```

MATLAB Script

```
function X=Iint(A,p,Y)
%using eqn 6.34
[n,N]=size(Y);
X=zeros(n,N);
Ai=inv(A+p*eye(n,n));
X(:,1)=Ai*Y(:,1);
alpha=-2*p*X(:,1);
for i=2:N;
X(:,i)=Ai*(Y(:,i)+alpha);
alpha=alpha-2*p*X(:,i);
end
```

```
MATLAB Script
function [Ap,L0]=lagc(p,N)
%Generating system matrix Ap
Ap=-p*eye(N,N);
for ii=1:N
for jj=1:N
if jj<ii, Ap(ii,jj)=-2*p;
end
end
end
L0=sqrt(2*p)*ones(N,1);
```

(10) Plot $\hat{\mathbf{y}}(t)$ and control input $\bar{\mathbf{u}}(t)$.

```
MATLAB Script
figure()
subplot(231)
plot(t1,u1(1,:))  %,'-',t,u_max(1,1).*ones(length(u1),1),'--')
ylabel('surge')
xlabel('time')
subplot(232)
plot(t1,u1(2,:))
ylabel('sway')
xlabel('time')
title('u - constrained w alpha')
subplot(233)
plot(t1,u1(3,:))
ylabel('heave')
xlabel('time')
subplot(234)
plot(t1,u1(4,:))
ylabel('roll')
xlabel('time')
subplot(235)
plot(t1,u1(5,:))
ylabel('pitch')
xlabel('time')
subplot(236)
plot(t1,u1(6,:))
ylabel('yaw')
xlabel('time')
```

```
figure()
subplot(231)
plot(t1,y1(1,:))
ylabel('surge')
xlabel('time')
subplot(232)
plot(t1,y1(2,:))
ylabel('sway')
xlabel('time')
title('output - constrained w alpha')
subplot(233)
plot(t1,y1(3,:))
ylabel('heave')
xlabel('time')
subplot(234)
plot(t1,y1(4,:))
ylabel('roll')
xlabel('time')
subplot(235)
plot(t1,y1(5,:))
ylabel('pitch')
xlabel('time')
subplot(236)
plot(t1,y1(6,:))
ylabel('yaw')
xlabel('time')
```

```
figure()
subplot(231)
plot(t1,udot1(1,:))
ylabel('surge')
xlabel('time')
subplot(232)
plot(t1,udot1(2,:))
ylabel('sway')
xlabel('time')
title('udot - constrained w alpha')
subplot(233)
plot(t1,udot1(3,:))
ylabel('heave')
xlabel('time')
subplot(234)
plot(t1,udot1(4,:))
ylabel('roll')
xlabel('time')
subplot(235)
plot(t1,udot1(5,:))
ylabel('pitch')
xlabel('time')
subplot(236)
plot(t1,udot1(6,:))
ylabel('yaw')
xlabel('time')
```

In Fig. 4.32, the velocity responses show that the decentralized MPC using Edmunds scaling is feasible. From the results, the decentralized MPC algorithm can be implemented with a less numerical error. However, the proposed decentralized MPC with Edmunds scaling can be quite sensitive to some control parameters. The control amplitude constraints are within the lower and upper limits of the control

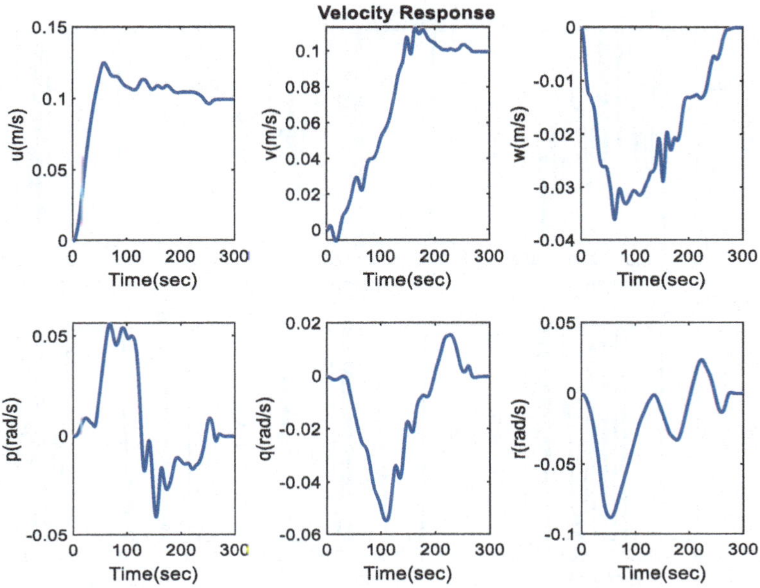

Fig. 4.32 Velocity response for decentralized model predictive control

amplitude, as can be seen in Fig. 4.33. In Fig. 4.34, the constraints on the rate of change in control amplitude that correspond to the changes in the thruster rate are below the maximum rate.

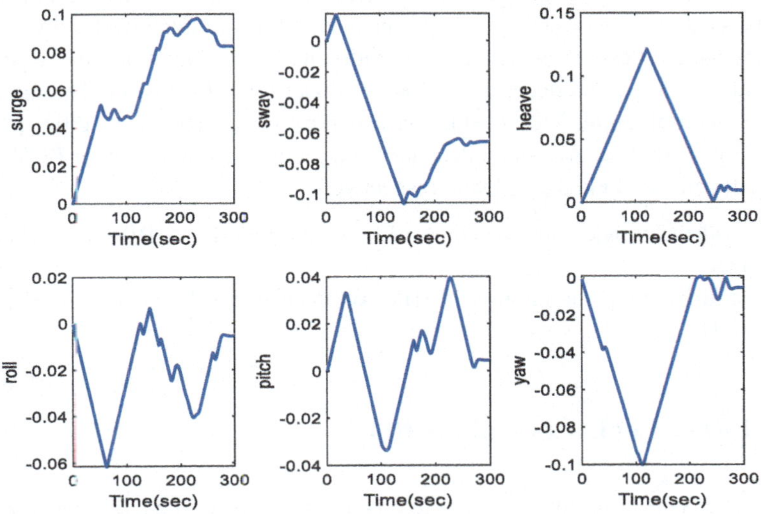

Fig. 4.33 Control inputs for decentralized model predictive control

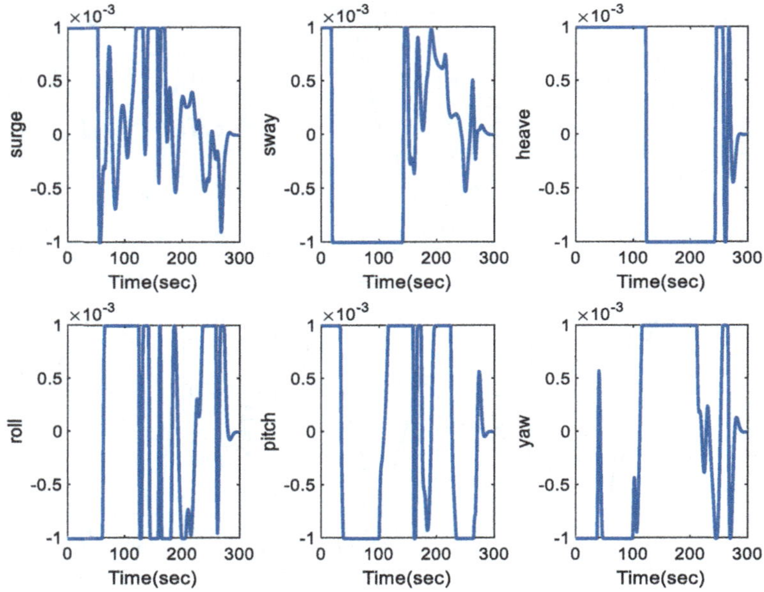

Fig. 4.34 Rate of control inputs for decentralized model predictive control

4.7 Summary

The robust decentralized model predictive control (MPC) of the ROV was proposed. The proposed Edmunds design scaling algorithm on the transfer function matrix equalizes and maximizes the geometric mean of the row dominance for given inputs and outputs. The design scaling makes the individual control loops more independent, hence able to achieve a decentralized ROV model for MPC layout. The numerical stability of the algebraic Riccati equation also improves with better condition number.

The linear ROV analysis and decentralized predictive controller for ROV model can be downloaded via the website, respectively.

- https://drive.google.com/open?id=1nEfnoGAhug9FGJOM5RjWHHW1hG9F8niB
- https://drive.google.com/open?id=1fV2WQKQEfqIDWKiEjesKQqYm1qzwVUH.

4.8 Fuzzy-Logic Control for ROV

There have been various efforts to develop the controller for the URV, which include both the (conventional) linear and the (modern) intelligent control schemes. Intelligent control methods which include neural networks (NN), sliding mode (SLC), and fuzzy-logic controllers (FLC) are more robust to hydrodynamics uncertainties.

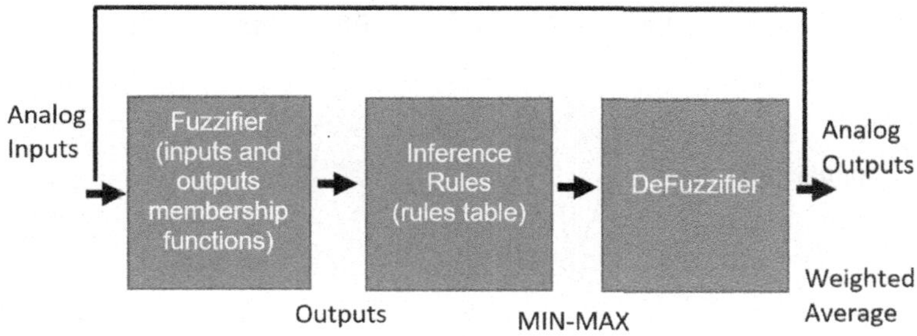

Fig. 4.35 Flow in fuzzy-logic controller design

Also, they exhibit excellent immunity to disturbances. Although intelligent control is very promising for URV application, it requires substantial computational power. For example, an FLC [1] has to deal with fuzzification, rule base storage, inference mechanism, and defuzzification operations. Despite these issues, it is known that an FLC has a simple control structure and offers a higher degree of freedom in tuning its control parameters.

The system under control is described in terms of some linguistic variables. The linear model of the ROV with Edmunds scaling is used. The control tasks are performed by using some rules outlined in a rule base. The inference rule used is a fuzzy inference scheme. As seen in Fig. 4.35, it takes three steps to design a fuzzy controller: fuzzification, inference rules, and defuzzification. In the first step, the values obtained through a sensor are transformed into values of the corresponding linguistic variable. The second step performs the fuzzy inference giving the linguistic values of the control variables. In the third step, these linguistic values are transformed into the numerical value of the control variable to perform the required task. The controller can be fine-tuned iteratively.

First, the FLC has two linguistic input variables or sometimes called controlled inputs, namely error (e) and the change of error ($\dot{e}$). Each of the error and change of error fuzzy sets has seven membership functions, as shown in Fig. 4.36. The triangular and trapezoidal membership functions for inputs have seven linguistic variables defined as negative big (NB), negative medium (NM), negative small (NS), zero (Z), positive small (PS), positive medium (PM), and positive big (PB). As observed, all linguistic variables have the same widths. PL and NL are the saturation limits imposed on the FLC.

On the other hand, the linguistic output variable is the voltage applied to the thrusters. The output membership functions have seven singleton functions as shown. It has the same linguistic variables as NB, NM, NS, Z, PS, PM, and PB. The singleton is represented by an individual point in the output space. Truncation of singletons results in the reduction of the input.

The fuzzy inference operation or process is implemented by using 49 rules (or if–then statements). The min–max (or implication–aggregation) compositional rule of inference and the weighted average method has been used in the defuzzifier process based on the Sugeno inference process.

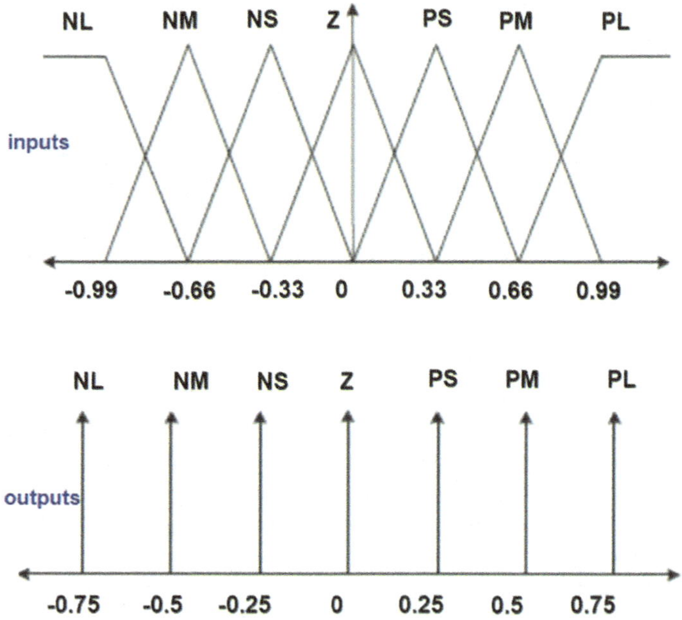

Fig. 4.36 Input and output membership functions for Sugeno type

Here the AND method truncates the output fuzzy set (that is the minimum of the linguistic variables). Conversely, if OR operator is used, it selects the maximum of the two variables. Similarly, the fuzzy inference process can be captured by the fuzzy interference diagram as shown in Fig. 4.37.

The 49 rules are created on a two-dimensional space of the phase-plane $(e, \dot{e})$. The rule table has the same output membership in a diagonal direction that is known as the Toeplitz structure. Now, how to model and simulate the fuzzy logic? The graphical

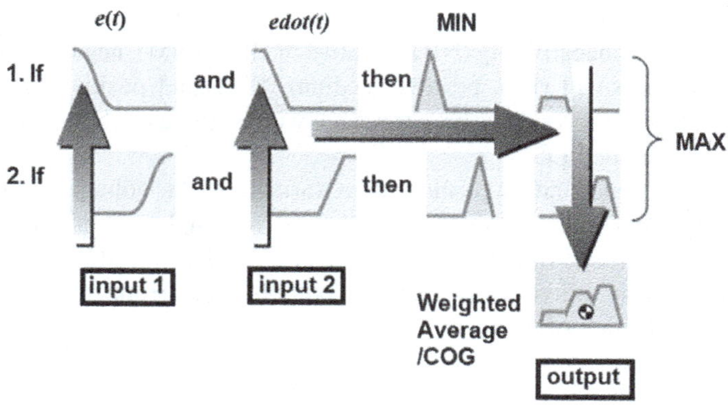

Fig. 4.37 Fuzzy interference diagram

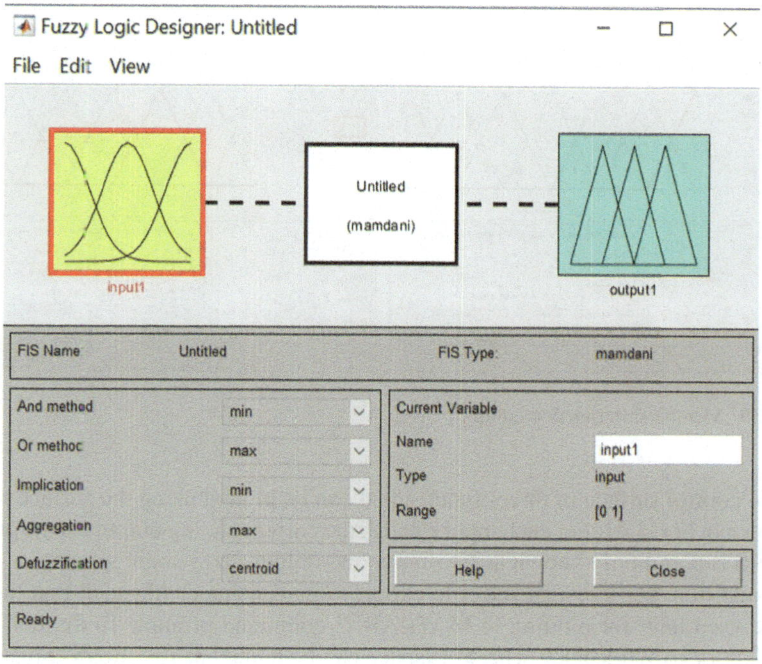

Fig. 4.38 FIS editor

user interface (GUI) tool provided by the fuzzy-logic toolbox is used. Although it is possible to use the fuzzy-logic toolbox by working strictly from the command line, in general, it's much easier to build a system graphically. There are few primary GUI tools for building, editing, and observing fuzzy inference systems in the fuzzy-logic toolbox.

First, type "fuzzy" after the MATLAB™ command prompt. The FIS Editor will pop up as shown in Fig. 4.38. Note that the Sugeno FIS was chosen.

The next step is to create the inputs membership functions for e and $\dot{e}$. It is followed by creating the output membership function for the control voltage to the thruster. In Fig. 4.29, the completed membership functions can be seen in the membership function editor for $\dot{e}$ and e, respectively (Fig. 4.39).

The FIS is not completed until the rules in Fig. 4.40 are implemented in the Rule editor.

As shown in Fig. 4.40, there are 49 rules or if–then statements. It displays a roadmap of the whole fuzzy inference process. You see a single figure window with 49 small plots nested in it. Each rule is a row of plots, and each column is a variable. The first two columns (the yellow plots) of plots show the membership functions referenced by the antecedent or the if-part of each rule. The third column of plots (the dark blue plots) shows the membership functions referenced by the consequent or the then-part of each rule.

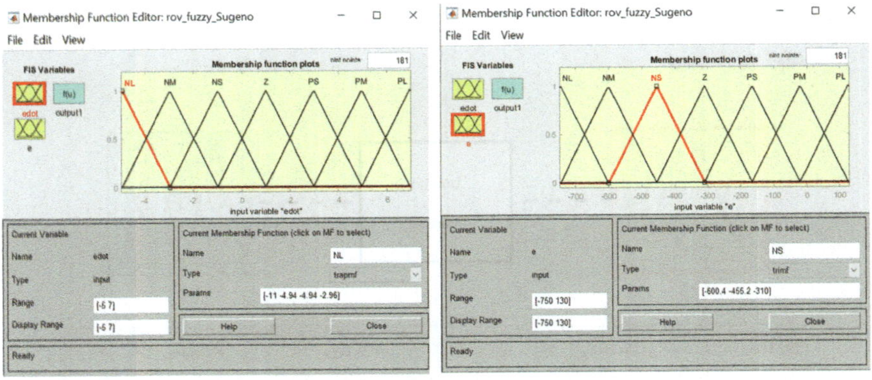

Fig. 4.39 Membership function editor

The control surface in three-dimensional can be plotted using the surface viewer as shown in Fig. 4.41. For each input e and $\dot{e}$, the corresponding output can be plotted. The FIS is commonly known as two-input one-output fuzzy-logic systems.

In addition, the user can select to export to workspace or file such that the data can be used later for plotting in MATLAB^{TM} command prompt. To do this, select File/Export/To Workspace. The workspace can be named as rov_fuzzy_Sugeno.fis. Besides using the GUI of the fuzzy-logic system and MATLAB^{TM} commands, the fuzzy-logic toolbox in Simulink^{TM} is used to simulate FLC as shown in Fig. 4.42. It can be observed that the PID controller is used to control the position in six degrees of freedom. The PID controller gains are as follows:

- Surge: $k_p = 50, k_i = 50, k_d = 50$;
- Sway: $k_p = 44, k_i = 450, k_d = 0$;
- Heave: $k_p = 10, k_i = 5500, k_d = 45$;
- Roll: $k_p = 10, k_i = 555, k_d = 10$;
- Pitch: $k_p = 10, k_i = 120, k_d = 10$;
- Yaw: $k_p = 40, k_i = 222, k_d = 45$;

In the simulation, the FLC system controls the ROV to the desired velocity ($x_d = y_d = 2$ m). It consists of a subsystem to compute the commanded voltage for the thrusters and a subsystem that contains the fuzzy-logic controller, as shown in Fig. 4.43. As seen in Fig. 4.44, there are six ports for each linguistic variable and each has a fuzzy-logic controller attached in series with it except for the last port that uses the PID controller (with the following gains: ($k_p = 20, k_i = 20, k_d = 1$).

The position and velocity-time responses can be seen in Figs. 4.45 and 4.46, respectively. As observed in the position plots, the surge position is quite oscillatory as compared to the sway position. The steady-state error for the surge position is quite high. However, both the surge and sway positions settle to steady-state after approximately 300 s. The surge and sway velocities reduce to zero as the positions reached the desired values. As observed, the roll and pitch angle is quite small.

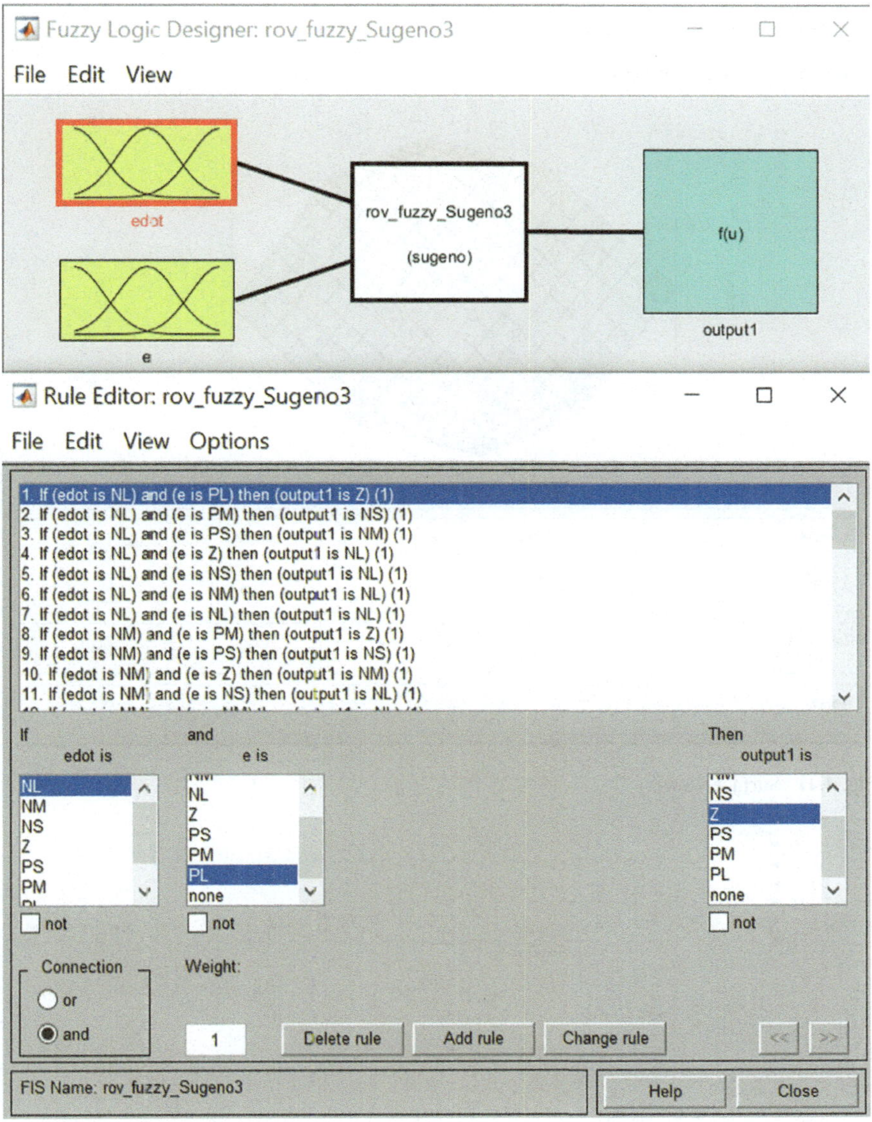

Fig. 4.40 Rule editor

Further improvement in the time responses can be made by fine-tuning the rules table values.

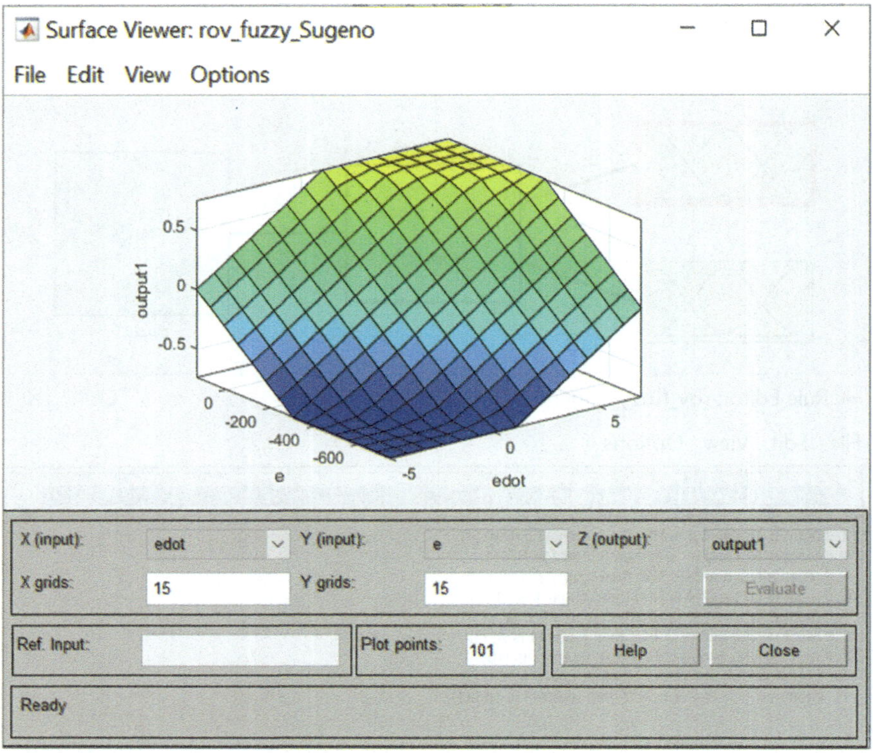

Fig. 4.41 Surface viewer

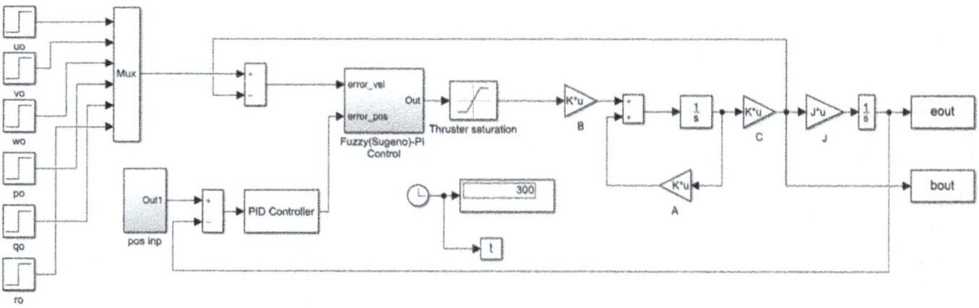

Fig. 4.42 Fuzzy-logic control for ROV with Edmunds scaling

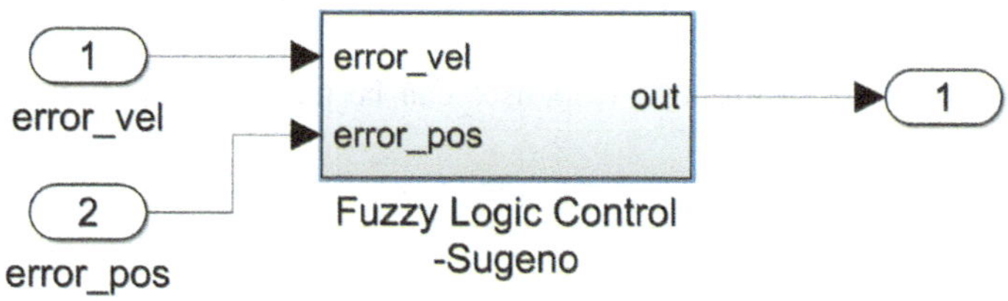

Fig. 4.43 Simulink block diagram for FLC

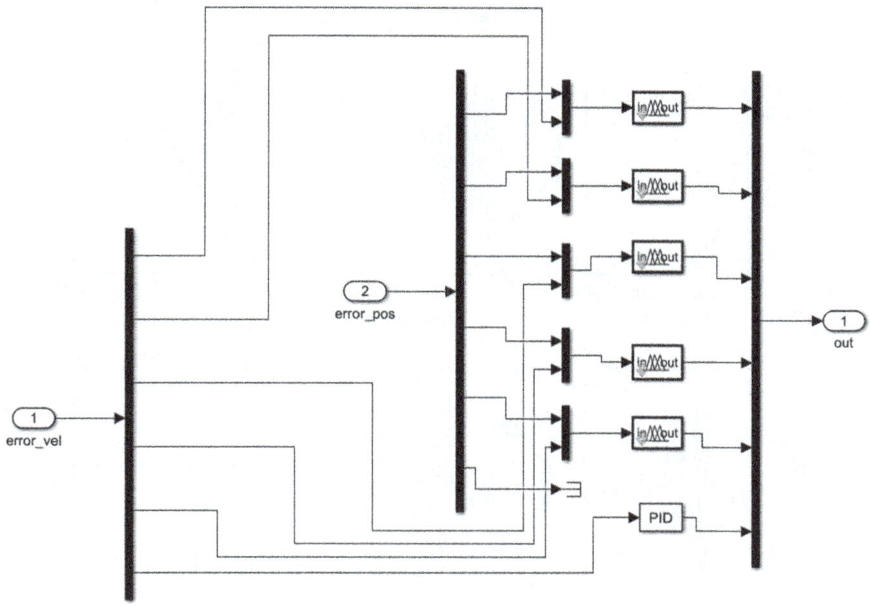

Fig. 4.44 Details of Simulink block diagram for FLC-Sugeno type

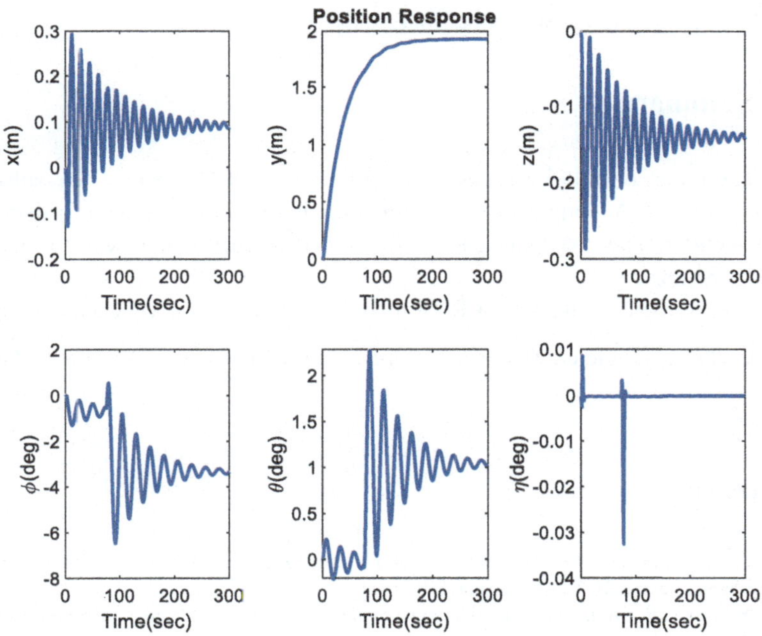

Fig. 4.45 Position time responses of FLC-Sugeno type

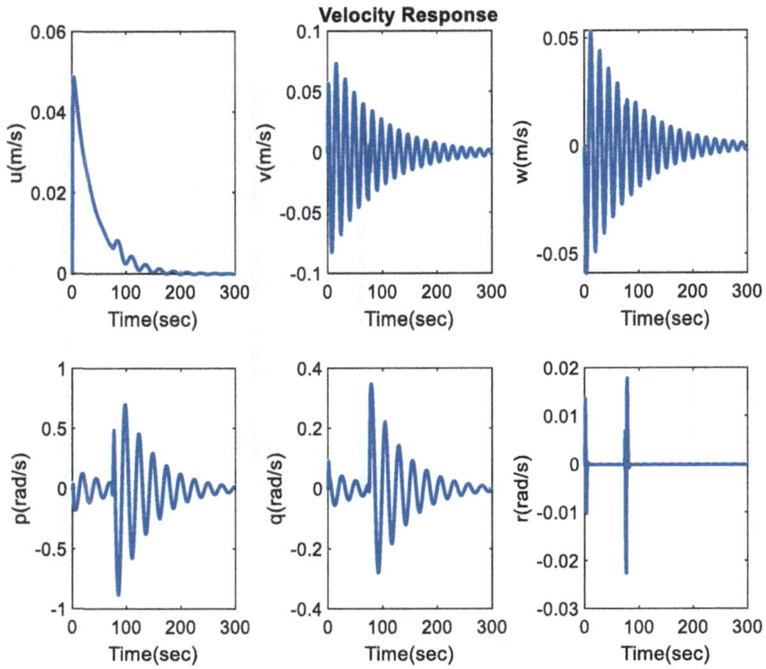

Fig. 4.46 Velocity time responses of FLC-Sugeno type

4.9 Summary

The fuzzy-logic controller was designed for the linear ROV model. The results have shown that the ROV could satisfy the time-domain performance with further needs for fine-tuning. The controller can be fine-tuned in an iterative way to reach the desired positions.

The fuzzy-logic controller for ROV model can be downloaded via the website:

- https://drive.google.com/open?id=1OpBlxw2ftaUhBBlEvz07YQiP3dwUaRdi.

References

1. C.S. Chin, *Computer-Aided Control Systems Design: Practical Applications Using MATLAB® and Simulink®* (CRC Press, Florida, 2017)
2. M. Eslami, C.S. Chin, A. Nobakhti, Robust modelling, sliding-mode controller and simulation of an underactuated ROV under parametric uncertainties and disturbance. J. Mar. Sci. Appl. **18**(2), 213–227 (2019)
3. C.S. Chin, *Embedded Mechatronics System Design for Uncertain Environments: Linux-based, MATLAB xPC Target, PIC, Arduino and Raspberry Pi Approaches* (IET Press, London, 2018)

4. C.S. Chin, W.P. Lin, Robust genetic algorithm and fuzzy inference mechanism embedded in sliding-mode controller for uncertain underwater robot. IEEE/ASME Trans. Mechatron. **32**(2), 655–666 (2018)
5. C.S. Chin, W.P. Lin, J.Y. Lin, Experimental validation of open-frame ROV model for virtual reality simulation and control. J. Mar. Sci. Technol. **23**(2), 267–287 (2018)
6. W.P. Lin, C.S. Chin, Block diagonal dominant remotely-operated vehicle model simulation using decentralized model predictive control. Adv. Mech. Eng. (SS: Advanced Intelligent Control of Autonomous Systems) **9**(4), 1–24 (2017)
7. M.D. Thekkedan, C.S. Chin, W.L. Woo, Virtual reality simulation of fuzzy-logic control during underwater dynamic positioning. J. Mar. Sci. Appl. **14**(1), 14–24 (2015)
8. W.P. Lin, C.S. Chin, L.C.W. Looi, J.J. Lim, E.M.E. Teh, Robust design of docking hoop for recovery of autonomous underwater vehicle with experimental results. Robotics **4**(4), 492–515 (2015)
9. D. Wan, C.S. Chin, Simulation and prototype testing of a low-cost ultrasonic distance measurement device in underwater. J. Mar. Sci. Technol. **20**(1), 142–154 (2015)
10. J.T. Si, C.S. Chin, An adaptable walking-skid for seabed ROV under strong current disturbance. J. Mar. Sci. Appl. **13**(3), 305–314 (2014)
11. C.S. Chin, M.S.W. Lau, Modeling and testing of hydrodynamic damping model for a complex-shaped remotely-operated vehicle for control. J. Mar. Sci. Appl. **11**(2), 150–163 (2012)
12. C.S. Chin, S.H. Lum, Rapid modeling and control systems prototyping of a marine robotic vehicle with model uncertainties using xPC target system. Ocean Eng. **38**(17–18), 2128–2141 (2011)
13. C.S. Chin, M.W.S. Lau, E, Low, Supervisory cascaded controllers design: experiment test on a remotely-operated vehicle. Proc. IMechE Part C: J. Mech. Eng. Sci. **225**(3), 584–603 (2010)
14. C.S. Chin, M.W.S. Lau, E. Low, G.G.L. Seet, Robust and decoupled cascaded control system of the underwater robotic vehicle for stabilization and pipeline tracking. Proc. IMechE Part I: J. Syst. Control Eng. **222**(4), 261–278 (2008)
15. Y.H. Eng, W.S. Lau, E. Low, G.G.L. Seet, C.S. Chin, Estimation of the hydrodynamic coefficients of an ROV using Free Decay Pendulum Motion. Eng. Lett. **16**(3), 326–331 (2008)
16. C.S. Chin, M.W.S. Lau, E. Low, G.G.L. Seet, Dynamic modelling and cascaded controller design of a low-speed maneuvering ROV. *Advanced Technologies: Research, Development and Application Book* (Advanced Robotic Systems International, Pro Literatur Verlag, 2006), pp. 159–186
17. C.S. Chin, M.W.S. Lau, E. Low, G.G.L. Seet, Software for modelling and simulation of a remotely operated vehicle. Int. J. Simul. Model. **5**(3), 114–125 (2006)
18. C.S. Chin, M.W.S. Lau, E. Low, G.G.L. Seet, Robust controller design method and stability analysis of an underactuated underwater vehicle. Int. J. Appl. Math. Comput. Sci. **16**(3), 101–112 (2006)
19. W.P. Lin, C.S. Chin, *Remote Underwater Dual Cameras Video Image Acquisition System using Linux Based Embedded PC104* (OCEANS'14 MTS/IEEE, Taipei, 2014), pp. 1–6
20. W. P. Lin, C.S. Chin. *Remote Robust Control and Simulation of Robot for Search and Rescue Mission in Water, OCEANS'14* (MTS/IEEE, Taipei, 2014), pp. 1–9
21. C.S. Chin, M.W.S. Lau, Y.J. Tan, K.F. Chee, Y.C. Wong, Development, and testing of an AUV using industrial xPC-Target platform, in *IEEE/ASME International Conference on Advanced Intelligent Mechatronics*, Singapore (2009)
22. C.S. Chin, M.W.S. Lau, E. Low, G.G.L. Seet, A cascaded nonlinear heading control with thrust allocation: an application on an underactuated ROV, in *IEEE International Conferences on Robotics, Automation and Mechatronic (RAM)*, Bangkok, Thailand (2006)
23. C.S. Chin, M.W.S. Lau, E. Low, G.G.L. Seet, Design of thrusters configuration and thrust allocation control for a remotely operated vehicle, in *IEEE International Conferences on Robotics, Automation and Mechatronic (RAM)*, Bangkok, Thailand (2006)

24. C.S. Chin, M.W.S. Lau, E. Low, G.G.L. Seet, Analysis and linear control of a low-speed maneuvering underwater robotic vehicle (URV), in *4th Asian Conference on Industrial Automation and Robotics*, Bangkok, Thailand (2005)
25. C.S. Chin, M.W.S. Lau, *Measurement and Control of a Thruster for Unmanned Robotic Vehicle (URV), 5th National Undergraduate Research Opportunities Programme* (NUROP) Congress, NTU, Singapore, 1999)
26. R. Lozano, B. Brogliato, Adaptive control of robot manipulators with flexible joints. IEEE Trans. Autom. Control **37**(2), 174–181 (1992)
27. P.V. Kokotovic, The joy of feedback: nonlinear and adaptive. IEEE Control Syst. Mag. **12**(3), 7–17 (1992)
28. C.E. Garcia, D.M. Prett, M. Morari, Model predictive control: theory and practice -a survey. Automatica **25**(3), 335–348 (1989)
29. W.H. Chen. D.J. Balance, J.O'Reilly, Model predictive control of nonlinear systems: computational burden and stability. IEE Proc. Part D: Control Theory Appl. **147**(4), 387–394 (2000)
30. P. Lu, Optimal predictive control for continuous nonlinear systems. Int. J. Control **62**(3), 633–649 (1995)
31. M. Soroush, H.M. Soroush, Input-output linearising nonlinear model predictive control. Int. J. Control **68**(6), 1449–1473 (1997)
32. J.M. Edmunds, Input and output scaling and reordering for diagonal dominance and block diagonal dominance. Proc. IEE Part D **145**, 523–530 (1998)
33. M.E. Salgado, A. Conley, MIMO interaction measure and controller structure selection. Int. J. Control **77**(4), 367–383 (2004)
34. L.P. Wang, *Model Predictive Control System Design and Implementation using MATLAB ®* (Springer, London, 2009)
35. D.A. Smallwood, L.L. Whitcomb, Model-based dynamic positioning of underwater robotic vehicles: theory and experiment. IEEE J. Oceanic Eng. **29**(1), 169–186 (2004)
36. R.K.A. Seyab, Y. Cao, S.H. Yang, Predictive control for the ALSTOM gasifier problem. IEE Proc. Control Theory Appl. **153**(3), 293–301 (2006)
37. J.C. Gomez JC, A. Jutan A, E. Baeyens, Wiener model identification and predictive control of a pH neutralisation process. IEE Proc.—Control Theory Appl. **151**(3), 329–338 (2004)
38. T.M. Howard, M. Pivtoraiko, R.A. Knepper, A. Kelly, Model-Predictive motion planning: several key developments for autonomous mobile robots. IEEE Robot. Autom. Mag. **21**(1), 64–73 (2014)
39. L.J. Zhang, X.T. Zhuan, Optimal operation of heavy-haul trains equipped with electronically controlled pneumatic brake systems using model predictive control methodology. IEEE Trans. Control Syst. Technol. **22**(1), 13–22 (2014)
40. D. Castaignet, I. Couchman, N.K. Poulsen, T. Buhl, J.J. Wedel-Heinen, Frequency-Weighted model predictive control of trailing edge flaps on a wind turbine blade. IEEE Trans. Control Syst. Technol. **21**(4), 1105–1116 (2013)
41. M. Preindl, E. Schaltz, P. Thogersen, Switching frequency reduction using model predictive direct current control for high-power voltage source inverters. IEEE Trans. Industr. Electron. **58**(7), 2826–2835 (2011)
42. W. Qi, J.F. Liu, P.D. Christofides, Distributed supervisory predictive control of distributed wind and solar energy systems. IEEE Trans. Control Syst. Technol. **21**(2), 504–512 (2013)
43. D.S. Alireza, D.A. Khaburi, R. Kennel, An improved FCS–MPC algorithm for an induction motor with an imposed optimized weighting factor. IEEE Trans. Power Electron. **27**(3), 1540–1551 (2012)
44. L.V. Steenson, S.R. Turnock, A.B. Phillips, C. Harris, M.E. Furlong, E. Rogers, L. Wang, Model predictive control of a hybrid autonomous underwater vehicle with experimental verification. Proc. Inst. Mech. Eng. Part M: Eng. Marit. Environ. **228**(2), 166–179 (2014)

45. T. Keviczky, F. Borrell, K. Fregene, D. Godbole, G.J. Balas, Decentralized receding horizon control and coordination of autonomous vehicle formations. IEEE Trans. Control Syst. Technol. **16**(1), 19–33 (2008)
46. C.S. Chin, N. Munro, Control of the ALSTOM gasifier benchmark problem using H2 methodology. J. Process Control 13(8), 759–768 (2003)

Conclusion

The book covers the modeling, identification, simulation, and control systems design of a benchmark model for a special class of underwater robotics vehicle (URV), namely, remotely operated vehicle (ROV). The benchmark model is useful for students, engineers, researchers, and lecturers who do not have a good reference model (with experimental validations) for the control systems design. Hence, this book serves as a useful guide to accomplish the simulation via computer-aided design using Simulink^{TM} as part of the MATLAB^{TM} software. Importantly, the MATLAB^{TM} programming codes and Simulink^{TM} block diagrams are provided for each chapter to allow users to practice and modify to suit their applications. The linear analysis and ROV controller can be downloaded via the following website.

- https://drive.google.com/open?id=1Vpy0k3ThbR2o67EGU3nxDu3Ewk9wX8ok.

In the modeling chapter, the dynamic and kinematics equations of the ROV were formulated. The mass inertia and buoyancy and gravitational force vector used in the dynamic equations, the hydrodynamic damping, and added mass components were derived. The simplified models for different planes and linear model of ROV were also defined.

In the simulation chapter, the dynamic and kinematics equations of the ROV were simulated using MATLAB^{TM}/Simulink^{TM}. The following block diagrams were modeled for the ROV:

- Mass inertial with added mass matrix;
- Coriolis and centripetal with added mass matrix
- Hydrodynamic damping force matrix
- Buoyancy and gravitational force matrix
- Euler transformation
- Thruster configuration matrix.

In the identifications of the benchmark ROV model, several techniques to estimate the model and the associated parameters for the analysis of the ROV performance and

C. S. Chin and M. W. S. Lau, *Benchmark Models of Control System Design for Remotely Operated Vehicles*, https://doi.org/10.1007/978-981-15-6511-3

model-based control of the vehicle were discussed. It is commonly known that the lab-based methods using physical model experiments are expensive. The computational methods using the computational fluid dynamics software such as STAR-CCM+^{TM} are available for another way of determining these parameters. The potential flow code using WAMIT^{TM} was used to predict the added mass on the ROV model obtained from MULTISURF^{TM} using the panel method to solve the potential flow around the vehicle. The simulated results were verified with the experimental tests in the water tank. The model and the values obtained were tested in the experimental trials in the water tank. The proposed method provides a viable alternative with reasonable results at an early design stage. It provides a fairly sufficient benchmark model and insight into the ROV behavior for controlling the ROV.

In the control systems design chapter, there are a few control systems that the authors have already published in the literature. They are namely:

- Nonlinear proportional–integral–derivative (PID) control with optimization
- Nonlinear sliding mode control
- Nonlinear velocity state feedback linearization
- Nonlinear fuzzy logic control
- Nonlinear cascaded system control
- LQG/LTR controller design
- H-Infinity controller design.

To further strengthen the topic, the control system design using PID-based cascaded controllers on the benchmark ROV model was simulated using Simulink^{TM}. The inner loop controller controls the velocity of the ROV while the outer controller moves the ROV to the desired X–Y positions. Also, the backstepping was used to design stabilizing controls for the benchmark ROV model. The responses showed more damped behavior than the PID controller due to the feedback-stabilizing control law of the backstepping controller.

Subsequently, the robust decentralized model predictive control (MPC) of the ROV was proposed. The proposed Edmunds design scaling algorithm on the transfer function matrix equalized and maximized the geometric mean of the row dominance for given inputs and outputs. The design scaling made the individual control loops more independent. The numerical stability of the algebraic Riccati equation in MPC also improved due to the lower condition number. Lastly, one of the intelligent controllers such as the fuzzy-logic controller was designed for the linear benchmark model of the ROV. The results have shown that the ROV could satisfy the time-domain performance. The fuzzy-logic controller can be fine-tuned in an iterative way to reach the desired positions.

In summary, this book provides a good reference book for modeling, identification, simulation, and control systems design of the ROV using MATLAB^{TM}/Simulink^{TM}. With the MATLAB^{TM} programming codes and Simulink^{TM} block diagrams provided, we hope to make this world a better place for a "deep" collaborative research.

The manufacturer's authorised representative in the EU is Springer Nature Customer Service Centre GmbH, Europaplatz 3, 69115 Heidelberg, Germany. If you have any concerns regarding our products, please contact ProductSafety@springernature.com

Printed and bound by CPI Group (UK) Ltd, Croydon, CR0 4YY
30/09/2026
02225515-0004